AF678494

Laser Cleaning in Conservation:

An Introduction

Edited by

Martin Cooper PhD BSc

BUTTERWORTH
HEINEMANN

SWANSEA INSTITUTE
OF HIGHER EDUCATION

LIBRARY AND LEARNING
SUPPORT SERVICES
114838.
CLASS NO: MPE
621.366 COO

Butterworth-Heinemann
Linacre House, Jordan Hill, Oxford OX2 8DP
22 Wildwood Avenue, Woburn, MA 01801-2041
A division of Reed Educational and Professional Publishing Ltd

A member of the Reed Elsevier plc group

OXFORD BOSTON JOHANNESBURG
MELBOURNE NEW DELHI SINGAPORE

First published 1998

© Martin Cooper 1998

All rights reserved. No part of this publication may be reproduced in any material form (including photocopying or storing in any medium by electronic means and whether or not transiently or incidentally to some other use of this publication) without the written permission of the copyright holder except in accordance with the provisions of the Copyright, Designs and Patents Act 1988 or under the terms of a licence issued by the Copyright Licensing Agency Ltd, 90 Tottenham Court Road, London, England W1P 9HE. Applications for the copyright holder's written permission to reproduce any part of this publication should be addressed to the publishers

British Library Cataloguing in Publication Data
A catalogue record for this book is available from the British Library

ISBN 0 7506 3117 1

Library of Congress Cataloguing in Publication Data
A catalogue record for this book is available from the Library of Congress

Composition by Scribe Design, Gillingham, Kent
Printed and bound in Great Britain by The Bath Press.

Contents

Preface

My first introduction to lasers as a potential cleaning tool for conservation came in 1972 when Dr John Asmus gave a demonstration of laser cleaning at the Victoria and Albert Museum (V&A). John had recently been working in Venice making holograms of decaying sculptures with a ruby laser. It was during this time that he met Kenneth Hempel (the V&A's chief sculpture conservator) who was cleaning pollution-damaged sculpture on behalf of the 'Venice in Peril' Fund. Seeing the problems that Ken was having in removing the pollution crusts with an airbrasive machine, John demonstrated how a pulse of energy from his laser could vaporize black pollution crusts, seemingly without damage to the marble beneath. Ken was so impressed with what the laser could do that he invited John to the V&A to demonstrate the use of the laser to the conservators there. In his demonstration John cleaned small areas on several types of stone and marble. Even more surprisingly, he showed how he could remove foxing from prints on paper, again without appearing to damage the substrate material.

Clearly, this technology was exciting and seemed to have great potential, but there were some fundamental problems. The only lasers available at this time were large, expensive and slow (several seconds elapsed between each pulse of energy). A further drawback for the conservator was that no suitable delivery system existed for directing the beam to the target area. At this time one could only set up the laser on a laboratory bench and fire at a prepared target. To move the beam over the surface of a sculpture freely (as one could with an airbrasive) was impossible. It was for these reasons that the airbrasive machine became the favoured cleaning tool on external stone sculpture for the next twenty years.

It was during the mid-1980s, when I had been working on a series of complex polychrome sculpture treatments, that I began to think again of lasers. The task of overpaint removal required many months of continuous work through a binocular microscope, using microsurgical scalpels to cut through the layers. I had gradually abandoned the use of chemicals for this work because I had seen through the microscope how difficult it was to control them. Most sculpture surfaces and most paint layers on sculpture are porous. Chemicals are readily absorbed by such surfaces and if they are allowed to penetrate the substrate will often cause salts to form and result in blooming at a later stage. Even when using gels and packs to hold the cleaning agent on the surface, one still has to wash away the residue with a neutralizing chemical.

The substitution of mechanical methods for chemical treatments on polychrome sculpture produced greater control, but there were still problems. Watching myself at work through the microscope I was aware of how easy it was to make errors with the scalpel when one's concentration failed or when layers were more brittle than anticipated. The use of the scalpel could also impart an artificial sheen to the paint that falsified its appearance. It was during these long hours of work with the microscope that I began to recognize the limitations of the human hand and the inadequacies of conventional technology. The regular use of the microscope for observation seemed to demand a cleaning system that could match it in terms of precision.

My misgivings about the problems of cleaning polychrome sculpture were also accompanied by doubts about the efficacy of the airbrasive machine for removing pollution crusts from deteriorated stone. By the mid 1980s airbrasive machines had undergone a series of developments that had turned them into compact and efficient tools. The tuning systems had become so delicate that one could work with air pressures of 5psi and fine aggregates like aluminium oxide, ground nutshell and calcium carbonate. Despite these refinements I felt that airbrasive technology had reached the limit of its development and that there were fundamental flaws in the system.

The airbrasive system works by delivering a continuous flow of abrasive particles, propelled by compressed air, at an encrustation which is slowly worn away. It is not chemically discriminating and makes no distinction between coatings and substrates other than their hardness. The only variation in the particle stream is provided by the conservator who has to judge when one layer will disappear and another be revealed. Although the conservator's skill and dexterity are an important element in any treatment, with the airbrasive it is all-important. Layers of corrosion do not usually detach themselves uniformly. It is therefore easy for the conservator to remove both the corrosion and original surface without noticing. A generation of conservators have also learned to loathe the airbrasive machine for the dust particles it generates. Whereas small artefacts in the laboratory can be cleaned in an extraction chamber, large sculptures cannot be easily contained. The conservator has to don a protective suit and air helmet in the hope that they will not breathe in the dust. Unfortunately the fine abrasive particles will penetrate even the finest filters and at the end of a session the conservator will be covered in dust and chewing grit.

For these reasons, and many more, I began to make enquiries about progress in the field of laser technology. My introduction to modern laser technology really came through my interest in three-dimensional recording of sculpture. I had long felt that our methods for recording and measuring sculpture were woefully inadequate. Most museum collections of sculpture are still recorded with black and white photography and measured with a tape rule. Such technology (readily available in the mid-nineteenth century) could not do justice to the microscopic precision that I required to record minute changes in surface on three dimensional objects. For several years I had worked with specialists in stereo-photography and also close-range photogrammetry. Although this produced some exciting images and allowed us to record surfaces in three dimensions, it did not give us the detail or accuracy that we required. Scientific colleagues suggested that laser scanners might provide the answer and so I became involved with the world of lasers and opto-electronics.

The inherent precision of the laser beam and its ability to minutely record the surface of an object when linked to a CCD camera, led me once again to think of using lasers for cleaning. I met Professor David Emmony at Loughborough University in 1986 and he undertook some cleaning experiments in his laser laboratory. The tests he carried out on polluted limestone and marble with a small neodymium:YAG (Nd:YAG) laser were so successful that we decided to institute a research programme into laser cleaning. When I moved from the V&A in 1991 to join the National Museums and Galleries on Merseyside (NMGM) I was greatly encouraged by the enthusiasm with which laser technology was greeted. Martin Cooper, who had studied laser cleaning for his PhD at Loughborough, joined me in Liverpool, as did Stephen Fowles who specialized in laser scanning. We have now formed a Laser Technology section in the Conservation Centre at Liverpool, which continues to research these aspects of laser technology. Our work during the past five years has been helped by generous grants from the Henry Moore Foundation, the Leverhulme Trust and the Headley and Monument Trusts. We have also developed a stimulating relationship with Lynton Lasers Ltd who are commercial manufacturers of medical laser systems and have helped to turn our prototype cleaning laser into a commercial product.

The main thrust of our research during the past five years has been to evaluate laser cleaning on behalf of the average conservator working either in museums or private practice. The initial cost of investing in laser technology is high but its flexibility and low running costs can make it economic in the long term. We have tested a wide range of materials with the Nd:YAG laser including marble, limestone, sandstone, stucco, plaster, aluminium, terracotta, ivory, vellum and even basketry, all with great success. This demonstrates the wide range of materials that this type of laser radiation will clean and its value to the conservator or a whole conservation department. In our research the Nd:YAG laser, an infrared laser, has been shown to have great benefits when cleaning organic materials. We have found that we can remove dirt layers from materials such as shellac, wax and vellum without swelling or overheating the substrate material. Coatings such as cellulose nitrate can be detached from the most fragile stonework.

One of the most important features of our research to emerge has been the need for practical testing with every new material. Theory and practice do not always coincide and we have often found that our theoretical predictions have been confounded. When we began our testing we assumed that removing dark material from light (e.g. pollution from marble) would be easy and the reverse virtually impossible. Although it is generally easier to remove dark materials with the laser, we have also found that we can remove materials such as limewash from brown sandstone without damaging the substrate. These results demonstrate that the reaction of laser energy with different materials is complex and often unpredictable. We have identified some of the mechanisms by which materials are removed from the substrate, but the detail of many interactions remains obscure.

For the conservator who learns to use a laser, many of the mechanisms we have identified can be used to good effect. For instance, the ablation mechanism where dirt layers heat up and are then ejected from the surface is a common effect on pollution encrustations. Vaporization also occurs on many materials, where the dirt is heated to a high temperature and turns to vapour. The ultrasonic action, where the laser energy can be used like a light chisel to separate one layer from another, is particularly useful when removing cement deposits from friable stone. These reactions can often be enhanced by the addition of water to the surface; future research will look into the use of other chemicals which when activated by laser energy can enhance cleaning.

One of the goals that I have always identified for laser technology is the removal of layers of overpaint on polychrome sculpture. I have always seen this as an exceedingly sophisticated problem and one that will not be easily solved. We have held back from looking into this problem, preferring to concentrate on those areas where we felt that we might achieve positive results and thus develop our understanding of the technology. We can only applaud the brave work of Professor Costas Fotakis and his team in Crete, who have concentrated on cleaning polychrome surfaces with ultraviolet lasers and have produced some interesting developments in this field. I think that their approach has been exciting and the analytical systems that have been developed to monitor their treatments will be of great benefit to all conservators and analysts. The interaction of laser energy with varnishes, pigments and paint media is, of course, a complex subject and it will be some years before the effect of these treatments can be assessed. Conservators are taught to regard light as an enemy and it will be hard to convince some professionals that they should at least look at laser technology as a viable alternative to other treatments. One cannot, however, dismiss the care with which Professor Fotakis and his team have approached their project and one would wish that many other conventional techniques had been subject to the same vigour.

For those of us for whom the laser is no longer a research tool but a technique we use on a daily basis, the future developments look very exciting. The fundamental cleanliness of the technique means that a whole range of optical systems can be designed to aid the conservator before, and during, cleaning. It will not be long before the cleaning beam is joined by a cluster of other beams passed through an optical fibre, or articulated arm, delivery system. Lasers are regularly used for analysis and work is already underway to provide the cleaning laser with a facility to analyse the vapour produced by the beam. Scanning heads could also be incorporated which will record the surface in three dimensions, giving valuable information about the thickness of layers removed and the surface morphology of the substrate beneath the dirt layer. These innovations will turn the laser cleaning machine into a piece of equipment that will greatly assist the conservator, allowing the cleaning of materials that were previously untreatable, in situations where sophisticated analytical and recording systems are not readily available.

This book aims to give the interested reader a basic grounding of the use of lasers for cleaning in conservation. It is based on practical experience gained during the past seven years, a period of time which has seen development of a commercial laser cleaning system through co-operation between scientists and conservators, and intensive testing by conservators on a wide range of works of art. The book includes a brief history of the subject, from the early days of John Asmus through to the modern day, an explanation of how a laser works and descriptions of the unique properties of laser radiation and how these can be used to selectively remove dirt layers from a delicate substrate. A series of case studies is also presented to give the reader a feeling for the kind of work which can be carried out using lasers. A separate chapter has been devoted to the laser cleaning of paintings since this is a relatively complex area, compared to the laser cleaning of sculpture. A different type of laser is used and on-line monitoring techniques are necessary to control the process. The book is divided into seven chapters:

Chapter 1, an *Introduction,* discusses the fundamental differences between laser cleaning and more conventional methods of cleaning and describes the history of laser cleaning in conservation, from the pioneering work of John Asmus in the early 1970s to the development of commercial cleaning systems in the early 1990s. A brief review of the published literature is presented which includes research on the use of lasers for cleaning stone, terracotta, textiles, paper, stained glass and paintings. This review does not, by any means, cover all research which has been conducted on laser cleaning as, unfortunately, much of this research has not been published yet. A lot of interesting work will be included in the proceedings of two international conferences held in 1995 and 1997, Lasers in the Conservation of Artworks (LACONA) I and II, which are due to be published soon by Mayer & Comp of Vienna, Austria.

Chapter 2 describes what a laser is and how a laser works. It begins by discussing the nature of light and the electromagnetic spectrum and goes on to describe the processes which lead to the generation of laser radiation. The properties of laser radiation important to cleaning and the different types of laser used in conservation are also described.

Chapter 3, *Removal of surface layers by laser radiation*, describes how the unique properties of laser radiation can be harnessed to remove material from the surface of a substrate in an extremely controlled manner. The interaction of a laser beam with a surface and the resulting effects are described. This chapter also includes a brief review of laser cleaning in industry, the development of which has proceeded along similar lines to that of laser cleaning in conservation. Some of the ideas developed in the microelectronics industry are used to describe some of the effects caused by the interaction of a laser beam with dirt layers on a work of art. Research into laser ablation in conservation is outlined and the mechanisms involved in the laser cleaning of artworks are described.

Chapter 4, *Practical laser cleaning with a Q-switched Nd:YAG laser*, describes some of the practical techniques used by conservators to optimize the quality of cleaning achieved with a laser operating in the near infrared. It concentrates on the Nd:YAG laser since this is the type of laser used most commonly in conservation studios. The advantages and disadvantages of laser cleaning at this wavelength are presented and the health and safety implications of the technique are outlined.

Chapter 5 presents a series of case studies which describe the cleaning of a wide range of artworks using a Nd:YAG laser. This chapter aims to give the reader an idea of the kind of work which can be undertaken with such a laser and is based on work carried out at the National Museums and Galleries on Merseyside during the past five years. It includes case studies on marble, limestone, sandstone, terracotta, plaster, aluminium, ivory, bone, parchment and wood.

Chapter 6 is devoted to the use of lasers in the conservation of painted artworks. It includes a description of the mechanisms involved in the cleaning process at ultraviolet wavelengths and an outline of the diagnostic techniques used to control the process. A case study is presented.

Finally, Chapter 7 discusses some of the possible future developments of this exciting field.

Acknowledgements

I would like to thank my colleagues in the Sculpture Conservation section at NMGM, in particular John Larson and Dr Stephen Fowles, for many helpful discussions, valuable contributions and their boundless enthusiasm for lasers! I would also like to express my appreciation to everyone else who has contributed to the development and introduction of laser techniques in conservation during the past seven years. Among others these include Professor David Emmony, the Department of Physics at Loughborough University, Lynton Lasers, Dr Ken Watkins, Cheryl Foster, Steve Aston and Mark Dalton (Liverpool University) and, of course, John Asmus whose pioneering work in this field was undertaken over twenty years ago! I would also like to thank the British Museum, Trinity Fine Art Ltd, the Vivat Trust and the Dean and Chapter of Lincoln for allowing photographs of their important objects to appear in this book. Much of this work has been financed by the Leverhulme Foundation and the Headley and Monument Trusts. I am extremely grateful to Dr Vassilis Zafiropulos and Professor Costas Fotakis of the FORTH Institute in Crete who have contributed an important chapter on paintings to this book. Much of the pioneering work on laser cleaning of paintings has been carried out at FORTH who played host to the first international conference on Lasers in the Conservation of Artworks (LACONA) in October 1995. Finally, I would like to thank my wife, Jill, for her help and encouragement during the course of writing this book.

Martin Cooper

The work described in Chapter 6 is partially supported by the 'Laser Technology in the Conservation of Artworks' project No. 640 of the EPET II Programme (Greece), from the European Union (EU) Structural Funds and by the Ultraviolet Laser Facility operating at FORTH under the TMR Program (DGXII, ERBFMGECT 950021) of the EU. We would also like to thank Lambda Physik and Quantel for providing the lasers for this work. We especially thank Michalis Doulgeridis and Anna Fostiridou from the National Gallery of Athens, Milko den Leeuw from the Atelier voor Restauratie and Research van Schilderijen, Rui Vilar from the Technical University of Lisbon, Helen Ioakimoglou and Akis Voyatzis from the Technical Education Centre, Athens and finally, last but not least, Demetrios Anglos, Vivi Tornari, Ionna Zergiotti, Stelios Couris, Costas Balas, Aristedis Petrakis and Yiota Spetsidou from the Laser Group at FORTH.

Vassilis Zafiropulos
Costas Fotakis

Contributors

Dr Martin Cooper
Research Scientist, Laser Technology, National Museums and Galleries on Merseyside, Conservation Centre, Whitechapel, Liverpool L1 6HZ, UK

Prof. Costas Fotakis
Foundation for Research and Technology-Hellas, Laser and Applications Division, PO Box 1527, 71110 Heraklion, Greece
Department of Physics, University of Crete, Heraklion, Greece

John Larson
Head of Sculpture and Inorganic Conservation, National Museums and Galleries on Merseyside, Conservation Centre, Whitechapel, Liverpool L1 6HZ, UK

Dr Vassilis Zafiropulos
Foundation for Research and Technology-Hellas, Laser and Applications Division, PO Box 1527, 71110 Heraklion, Greece.

1 Introduction

Martin Cooper

The origins of the laser can be traced back to 1917 when Albert Einstein published an article in which he developed the fundamental ideas of emission and absorption of light (Einstein, 1917). These ideas provided the basis for development of the laser but it was not until 1960 that the laser was first demonstrated (Maiman, 1960). Since then development has been extremely rapid and the unique properties of laser radiation have led to applications in a wide range of fields. In industry lasers are used to cut, drill, weld and engrave materials, ranging from lace (Jackson, 1994) to steel (Gagliano and Zaleckas, 1972). In medicine lasers are used widely in ophthalmology and dermatology (Goldman, 1971), while the many military applications include missile guidance systems. Lasers are used in surveying, as a light source in three-dimensional scanning of objects and to produce objects in resin and other materials by advanced manufacturing techniques such as stereolithography. Applications of lasers are not restricted only to specialist fields but are also found in everyday life: lasers are used to read compact discs and bar-codes at supermarket checkouts, to generate holograms and to create exciting displays of light at concerts. In cleaning and removing surface layers, from micron-sized particles on semiconductors in the microelectronics industry (Imen, Lee and Allen, 1991; Engelsberg, 1995) to old paint layers on aircraft (Schweizer, 1995), applications have also been found. It is in this area that lasers are becoming widely used in conservation.

Laser radiation is an extremely intense form of monochromatic light which is emitted in a highly collimated beam. Its monochromatic nature means that the laser beam will interact very intensely with some materials but hardly at all with others, while its highly directional nature means that guiding the beam accurately to a target is relatively straightforward. The directionality of the laser beam also means that its energy can be easily collected by a lens and either spread out over a large area or focused into a small spot to control its effect on a surface. It is these properties which make the laser so useful in such a wide range of applications.

Lasers exist in many shapes and forms. It is the choice of parameters such as wavelength and power output which is important when selecting a laser for a particular application. Lasers emit light at a wavelength which is characteristic of the lasing medium; this light can be in the ultraviolet, visible or infrared parts of the electromagnetic spectrum and is emitted either as a continuous beam or as a pulse. By correctly selecting the type of light, the amount of light and the manner in which the light is emitted the interaction of a laser beam with a material can be controlled precisely. Take, for example, manufacturing where lasers are used in processes as diverse as welding, engraving and tool alignment. In welding a high power laser acts as a controlled source of heat and continuous or long pulse (where the energy is deposited over a period of 1–10 ms) infrared lasers are used to melt the material to a specified depth without affecting the surrounding area. Engraving is usually carried out with high power short pulse (10–100 ns) infrared lasers. The laser is again acting as a controlled source of heat, but in this case the heating effect is so rapid that a specified amount of material is actually removed from the surface through vaporization. Low power continuous lasers operating in the visible part of the spectrum are often used in alignment applications. Since light travels in a straight line through a transparent liquid or gas a laser beam can be used in a similar way to a string or chalk line in the construction industry. Tunnels and pipelines are aligned using a laser beam, and it is now common practice to align industrial machinery using laser systems.

The laser is an extremely precise tool and selection of the most appropriate type for a particular application allows work to be carried out in a very accurate and controlled manner. The laser, therefore, has enormous potential in conservation where it is essential that treatment is carried out in this way. Cleaning often involves the removal of layers of dirt strongly bound to a weak and crumbling but valuable surface. Conventional cleaning techniques, such as micro-abrasive or steam cleaning, often prove successful in removing the dirt in such cases but not without causing some sort of damage to the underlying surface, either on a microscopic or even macroscopic scale. The damage is usually a direct result of the inability of the cleaning medium, be it abrasive particles or water molecules, to discriminate between what is to be removed from the surface and what is to be left intact. Particles of abrasive, for example, continue to cut away at a fragile surface once the layers of dirt have been removed. Careful use of a suitable laser system, however, allows the conservator to remove layers of dirt in a much more controlled manner so that the underlying surface can be maintained intact. Instead of bombarding a surface with abrasive particles or water molecules, a pulse of light is used so that any impact on the surface is negligible – the process is essentially non-contact. The fundamental difference between the laser and other forms of cleaning is that the laser beam is, in many cases, able to discriminate far more successfully between a clean and a dirty surface. This is a direct result of the monochromatic nature of laser radiation. Since the degree to which laser radiation interacts with a material depends to a large extent on its wavelength, a laser can be chosen so that interaction of the beam is much more intense with the layers of dirt on an object's surface than with the object itself. Once the layers of dirt have been removed the cleaning process is halted as the laser beam is harmlessly reflected away and the surface, whatever its condition, is revealed without suffering further damage.

The laser is by no means a panacea. Just like any tool there are problems to which it is ideally suited and problems to which it is not suited at all. If the laser is used inappropriately or carelessly it can cause damage to an object, but if it is used carefully by a trained conservator then it offers precision and control not previously possible using more conventional techniques. Since the late 1980s use of the laser in conservation has become more widespread as understanding of the processes involved has improved and the technology developed. Commercial laser systems, which tend to employ a pulsed infrared laser, are now available and are used on a routine basis in a number of conservation studios in the United Kingdom, France and Italy, where most of the research and development has been based.

The need for refinement of cleaning techniques

Following an initial assessment the first stage in the conservation of an object is often cleaning. Removal of an unsightly, black pollution encrustation from a crumbling medieval sculpture, for example, is not only desirable from an aesthetic point of view but also serves to reveal its true condition. Damage previously obscured by the encrustation can be treated and its cause located and eliminated, thereby arresting the decay and extending the life of the sculpture. Cleaning is clearly one of the

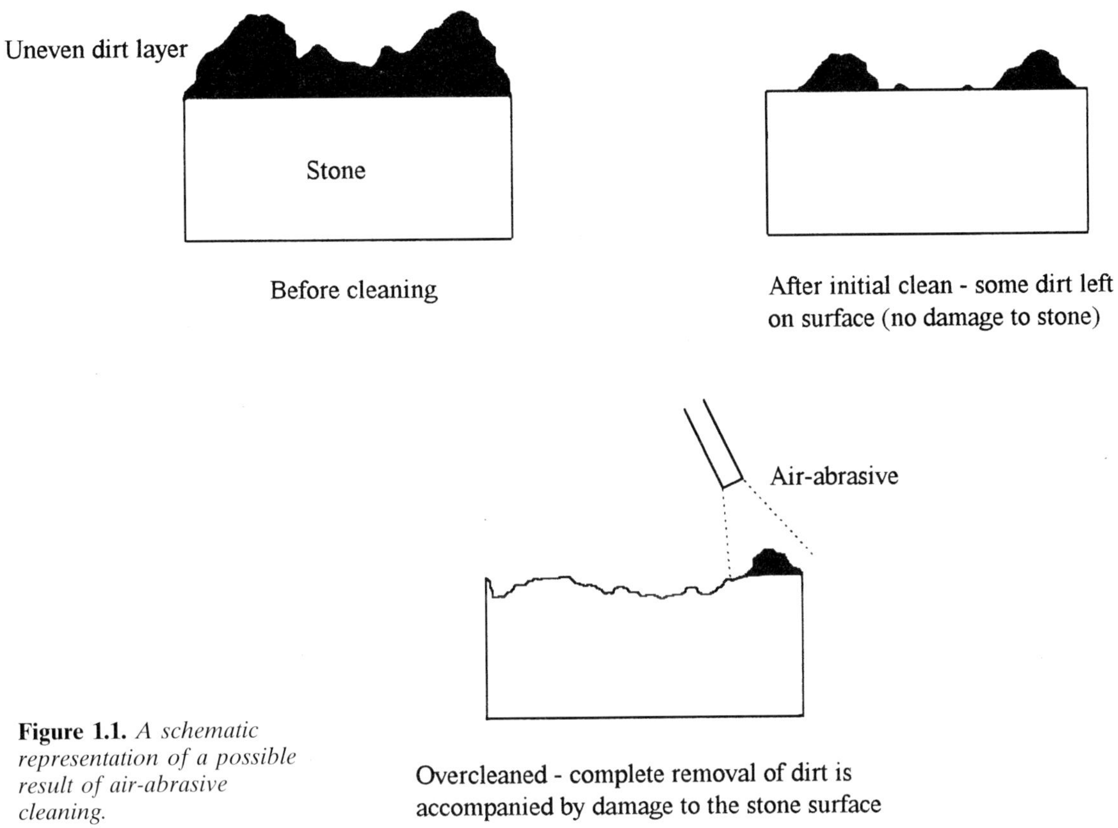

Figure 1.1. *A schematic representation of a possible result of air-abrasive cleaning.*

most important stages in the conservation process. If carried out carelessly it can worsen the appearance of the object or cause further damage which might even lead to acceleration of the decay process. With pollution levels as high as they are today objects, particularly outdoor sculpture in urban areas, are being cleaned with increasing frequency. It is, therefore, more important than ever before that cleaning is carried out by skilled conservators using the most refined techniques available.

Conventional methods of cleaning are often successful in restoring the outer appearance of an object, but since they all involve some form of contact with the surface it is not possible to clean an object completely without causing some kind of damage. At best this damage will be on a microscopic scale invisible to the naked eye, at worst it will be glaringly obvious, not only detracting from the appearance of the object but also increasing its susceptibility to further decay. Mechanical methods of cleaning involve collisions between the layers of dirt and a cleaning medium, the aim of the collisions being first to break the contact between the dirt and object surface and second to remove the dirt from the object. Once the dirt has been removed collisions between the object surface and cleaning medium continue until the conservator stops the process. This might only be for a small fraction of a second but, since the surface is often in a fragile state, will result in the irreversible loss of original material.

One of the most common mechanical methods is micro-abrasive cleaning which uses microscopic particles in a stream of compressed air as the cleaning medium. In order to clean a surface the abrasive particles must be harder than the dirt particles so that on impact the sharp edges are able to cut away the dirt layer. However, it is often the case that the dirt layer is harder than the underlying object surface and since the abrasive is unable to discriminate between the different materials the cutting action will continue until the abrasive is directed elsewhere. A black pollution encrustation, for example, is typically much harder than a block of weathered limestone, the surface of which may have reacted chemically to form a delicate patina. In this case abrasive cleaning, no matter how skilfully carried out, will result in some loss of material from the patina if the encrustation is completely removed. In extreme cases surface detail from a carving can be lost and soft, crumbling stone exposed to further decay processes. A typical result of abrasive cleaning is shown schematically in Figure 1.1. Dirt is easily removed from some areas but remains where it has formed a harder crust. In an attempt to remove these more stubborn deposits the abrasive continues to scour the surrounding area and complete removal of the dirt will produce an uneven surface made up of peaks and troughs. This is illustrated in Figure 1.2, which shows a polluted medieval limestone figure from York Minster. Where abrasive cleaning has been carried out very lightly much of the pollution crust remains and the surface still appears dirty. More thorough cleaning has exposed much more of the

(a) (b)

Figure 1.2. *Micro-abrasive cleaning tests on a limestone figure. (a) A life-sized medieval figure from York Minster. (b) Cleaning test areas (approximately 7.5 cm square). The lower square has been lightly cleaned and retains much of the pollution crust. The upper square has been more thoroughly cleaned, but appears 'speckled' due to the uneven nature of the pollution crust. Further cleaning would lead to some loss of original material.*

stonework, but the surface has adopted a 'speckled' appearance as a result of the uneven nature of the pollution crust. Complete removal of the crust would lead to overcleaning of some areas and loss of original material.

Water-based cleaning methods are often used to clean stone; water is cheap, readily available, non-hazardous to the conservator and the strong polarity of its molecules means that it is effective in removing a wide range of pollutants. Water can be applied either as a jet or as a fine mist and is often accompanied by scrubbing with a brush to remove the more stubborn deposits of dirt. However, water penetration can be a problem, leading to freeze-thaw damage, staining and mobilization of soluble salts, and the mechanical action of brushing can lead to damage of the object surface. A common water-based technique is steam cleaning in which steam is directed under pressure at the surface by a lance. The steam pressure is controlled by the conservator. Dirt is either dissolved, softened and removed by brushing or simply removed by the physical action of the steam. Steam cleaning should therefore only be used where the surface is in sound condition so that water cannot penetrate into the body of the object and the pressure of the steam is unable physically to disrupt its surface. Penetration of water into terracotta, for example, can give rise to startling changes in its appearance. Transportation of soluble salts from within the body of the material to the surface and subsequent crystallization as the water evaporates leads to the formation of unsightly patches and margins of white salts. This efflorescence, although reducing its visual appeal, is not damaging to the terracotta, but if conditions are such that evaporation takes place beneath the surface then the pressure exerted within pores by salt crystallization can be extremely damaging. An uncharacteristically deep red colour has been observed on some terracotta surfaces which have been steam cleaned (Plate 1.1), a phenomenon which has been attributed to enrichment of the surface iron concentration brought about by water penetration and subsequent movement of moisture back to the surface (Larson and Cooper, 1996).

The fundamental problems with water-based cleaning techniques are:

1. the physical action, whether the pressure at which water is delivered or the brushing, often results in some damage to delicate surfaces;
2. once the water has entered into the body of the material its effect on the object can no longer be controlled and its presence can lead to problems long after completion of cleaning, and;
3. the surface layer might be soluble.

If marble sculpture is exposed to a polluted atmosphere then areas of its surface which are sheltered from direct rainfall undergo a chemical transformation from calcite to the much more soluble gypsum (Verges-Belmin, 1994; Skoulikidis and Papakonstantinou-Ziotis, 1981; Lal Gauri and Holdren, 1981), the so-called patina, which is very difficult to retain if water-based methods are used to clean the sculpture.

Another important group of cleaning methods is based on the use of chemicals. The fundamental idea is that the selected chemical agent reacts with and dissolves the dirt layer without affecting the object surface. On a small scale the cleaning of stone surfaces using solvents and reagents applied by a cotton wool swab has been successful (Ashurst and Dimes, 1990), but on a larger scale there are many more risks, both to the object and to the conservator. Acids and acid salts are sometimes used to clean stone. The area to be cleaned is usually wetted prior to acid treatment to provide a damp surface on which the chemical will spread. The acid is then applied to the surface, either by brush, using a low pressure spray or as a gel, and is thoroughly washed off using a low volume water lance after the appropriate length of time. Acid cleaning is quiet, can be extremely effective in removing layers of dirt and often appears promising in the short term. However, problems can arise in the longer term and in some cases have led to accelerated weathering (Bluck and Porter, 1991a). Staining and bleaching can result from chemical reactions within the stone in which unusually large quantities of a substance are released to the surface, leaving a depleted region behind. Efflorescence and salt disturbance are direct results of the application of chemicals which react to produce salts, either by dissolving minerals within the stone and reprecipitating them on or close to the surface, or by reaction between two applied chemicals (sometimes an acid and alkali are used together so that one will neutralize the other). Strong acids not only remove unwanted layers of dirt but also etch the surfaces of minerals in the stone. This leads to increased water retention, encouraging the growth of unsightly and potentially damaging organic matter such as algae. These effects have all been observed in Glasgow, where hydrofluoric acid was used to clean a number of sandstone buildings. Studies carried out in Glasgow since (Bluck and Porter, 1991a; 1991b) have concluded that 'chemically cleaned buildings within the city are, at the moment, efflorescing, staining, decaying and soiling at a faster rate than those cleaned by mechanical methods'. Other disadvantages of using chemical agents include the inability to control their movement within an object and the inability to monitor their reaction once they have been applied. The conservator is unable to assess to what extent a fluid has penetrated an object, to halt aggressive reactions, or to prevent the fluid from migrating to areas which may be susceptible to damage. The conservator is also often unable to assess visually the reaction at the interface between dirt and object until the reagent has been washed off, by which time the object may already have been damaged irreversibly.

Poulticing is one technique that is used to avoid deep penetration and to limit the action of chemicals on an object. Chemicals are mixed with absorbent powders to form a paste or poultice which is applied to the surface of the object. As the poultice dries dirt is drawn out from the object and after several hours the 'mud' begins to fall away from the surface. The remainder of the poultice is then removed and the surface of the object cleaned with swabs and small quantities of de-ionized water. One of the most common poultices used in stone cleaning is made from de-ionized water and sepiolite (magnesium silicate). Using a poultice to clean stone has the advantage of wetting the surface in such a way that penetration of water into the stone is minimized and disruption of salts and minerals does not present such a problem. However, poulticing does involve physical contact with the object and the problem of being unable to monitor the cleaning whilst the poultice is in place persists.

Each of these methods of cleaning (mechanical, water and chemical-based) has advantages and disadvantages. In the hands of a skilled and experienced conservator, who understands the nature of the materials involved and the principles being used, the most appropriate technique is often successful in removing layers of dirt and restoring the appearance of an object. However, since each of these methods involves some form of intervention with the object, either mechanical or chemical, it is not possible to remove layers of dirt completely from an object without bringing about a change to its surface, however subtle. For example, it is not possible to remove dirt layers from stone using air-abrasive cleaning without causing collisions between abrasive particles and the stone surface, and the application of any solvent or reagent causes chemical collisions with both the dirt and stone. Monitoring the cleaning process poses a significant problem. The true state of the surface being cleaned is often obscured, either by abrasive powder, water or chemicals, and in some cases only becomes apparent after damage has been done. Instant visual feedback from the object would greatly reduce the risk of damage during cleaning. These fundamental problems mean that there will always be objects which are too fragile to be cleaned at all by 'conventional' techniques and objects which cannot be cleaned to the desired level for fear of damaging the surface. Clearly a refinement of cleaning methods is necessary so that once the conservator has identified what is to be removed from an object they can control the quantity of material removed, and from exactly where it is removed, in such a way that its surface is left unaltered. The availability of a technique which offers the control and precision necessary to achieve this would mean that the condition of the object after cleaning would be determined principally by the conservator and not by the method of cleaning selected.

Lasers offer a different approach to the problem of cleaning valuable and delicate objects. They have a number of advantages over more traditional methods:

1. *Non-contact* since the energy is delivered as light there is negligible contact with the object allowing extremely fragile surfaces to be worked on.
2. *Selectivity:* laser radiation is essentially monochromatic. By choosing a wavelength which is absorbed strongly by the dirt but only very weakly by the underlying surface it is usually possible to clean an object selectively, i.e. to remove the dirt layers without removing any material from the surface of the object. In effect the laser beam is able to discriminate between the dirt and the object. Such control means that the conservator is able to select exactly what is removed from a surface and allows him or her to go back over an area which has already been cleaned and remove any remnants of dirt without overcleaning. In this way it is possible to preserve the surface relief; original tool markings can be uncovered and delicate patinas left intact.
3. *Localized action:* the laser cleans only where directed. A single laser can supply a beam the size (diameter) of which is variable between a fraction of a millimetre and one centimetre, so the same tool can be used for both extremely precise and large scale work.
4. *Immediate control:* the cleaning action is instantly halted once the laser is switched off so the conservator can stop the process whenever he or she decides.
5. *Immediate feedback:* the condition of the surface can be continuously monitored by the conservator during cleaning. This allows decisions to be made much earlier during the process thereby reducing the risk of irreversibly damaging the object.
6. *Environmental:* laser cleaning does not generate large quantities of waste material. The only waste generated is the material removed from the object and this is relatively straightforward to collect and dispose of using efficient extraction systems. There is no use of hazardous chemicals or solvents and the only protective clothing necessary are safety spectacles and a face mask.

For these reasons laser cleaning is now becoming an accepted and important technique in conservation, offering control and precision not previously available to the conservator.

A brief history of laser cleaning

The early years: research conducted by John Asmus and colleagues

In 1972 a group of researchers from the University of California undertook a feasibility study in Venice to determine whether holography had a role to play in the conservation of stone sculpture (Asmus, Guattari, Lazzarini, *et al.*, 1973). During the first three months a pulsed ruby laser was used in-situ to make archival holograms of crumbling Venetian sculpture. The success of these holograms demonstrated a means of 'preserving' works of art in three dimensions. The collaboration between laser scientists and conservation scientists during this work led to the discovery that a pulsed ruby laser could also be used to remove black encrustations from decaying marble sculpture without apparent alteration to the marble surface. The idea of selectively vaporizing an optically absorbing substance from a reflective surface was not new and can be traced back to the 1960s when Arthur Schawlow, one of the pioneers of the laser, proposed a tool called the 'laser eraser' which would be able to selectively vaporize strongly absorbing black pigment from strongly reflecting white paper. John Asmus and co-workers suggested that a similar principle applied to the removal of black encrustations from white marble (Asmus, Murphy and Munk, 1973; Lazzarini and Asmus, 1973): the strongly absorbing black encrustation is removed by several pulses of laser radiation whereas the strongly reflecting white marble surface is left intact as the energy in the laser beam is simply reflected away. Asmus believed that if this was indeed the case in practice then the use of laser radiation would lead to a major advancement in cleaning techniques: a tool that could effectively detect a difference between layers of dirt and an object surface and respond accordingly, in a way that other techniques available at the time could not.

The potential of the technique was clear at a time when pollution levels were high and objects were being cleaned ever more frequently. The availability of a method which would allow a skilled conservator to remove thick black encrustations from a crumbling stone surface whilst maintaining a delicate and important patina intact was an exciting prospect. Asmus spent the next few years studying the effects of laser radiation on different materials and comparing the quality of cleaning achieved using a laser with that achieved using the more traditional techniques available at the time. Asmus proposed mechanisms, which have since been disputed and which have led to much debate between researchers (see Chapter 3), by which the cleaning action takes place; and developed a simplified theoretical model to aid understanding of the process. As a result of this work a prototype laser cleaning tool was developed and several cleaning projects were undertaken in Italy during the 1970s.

In 1973 Asmus and co-workers published a paper (Asmus, Murphy and Munk, 1973) describing initial work in which a pulsed ruby laser was used to remove dark encrustations from Venetian marble (Carrara) sculpture. It was observed that approximately three pulses were necessary to remove an area of encrustation (less than 0.5 cm thick) and expose the patina of the marble surface. Asmus also noticed that if the correct laser parameters were used then there appeared to be no alteration to the cleaned marble surface, even after repeated exposure to pulses once the encrustation had been removed. He argued that, provided the pulse length is short or the thermal conductivity of the material high, many materials with low optical absorption, i.e. high reflectivity or high transmissivity, will remain unaltered physically simply because little energy is actually deposited in the material. Tests were conducted using two different pulse lengths: 'normal-mode' (approximately 1 ms) and 'Q-switched' (approximately 35 ns), the origins of which are discussed later. The effect of the laser beam on the stone surface was found to be dependent on the pulse length, that is, dependent on whether the energy in the laser pulse is deposited into the surface over relatively long or short periods (e.g. milliseconds or nanoseconds). Asmus stated 'in general it was found that with normal-mode laser radiation no change in the surface was observed with repeated pulses after the dark material was removed' but that 'in some instances Q-switched pulses would continue to remove marble below the dark material'. These differences were explained on a qualitative basis:

- Removal of strongly absorbing material from a surface by normal-mode laser radiation occurs through thermal effects. Asmus argued that the temperature of an absorbing surface, such as black soot, is raised to 4000–5000 °K very quickly, resulting in vaporization. Once all of the absorbing material has been removed, further pulses are simply reflected away (see Figure 1.3) and, furthermore, since these processes occur very rapidly (milliseconds) 'very little heat is conducted into the base material (marble) and little or no damage occurs'.
- Removal of strongly absorbing material from a surface by Q-switched laser radiation occurs via different, more complex processes since the energy is deposited into the surface much more quickly (the surface temperature rise is therefore much higher and more rapid). Asmus described the removal process as consisting essentially of two parts (see Figure 1.4); vaporization of material at the beginning of the laser pulse, leading to formation of a strongly absorbing plasma (ionized gas) just above the surface, followed by compression and relaxation of the surface as the plasma first expands and then finally disappears when the laser pulse finishes. He argued that the second part of this process could result in removal of material from the marble surface as well as from the strongly absorbing encrustation above.

Asmus suggested that both of these cleaning modes have advantages and may be useful in different situations. The Q-switched mode was found to be more efficient in removing material and since the pulse length is so short the likelihood of thermally-induced microcracking in the marble below was thought to be even less than in normal-mode. Asmus noted however that 'on the other hand, the normal-mode pulses will not remove underlying light-colored

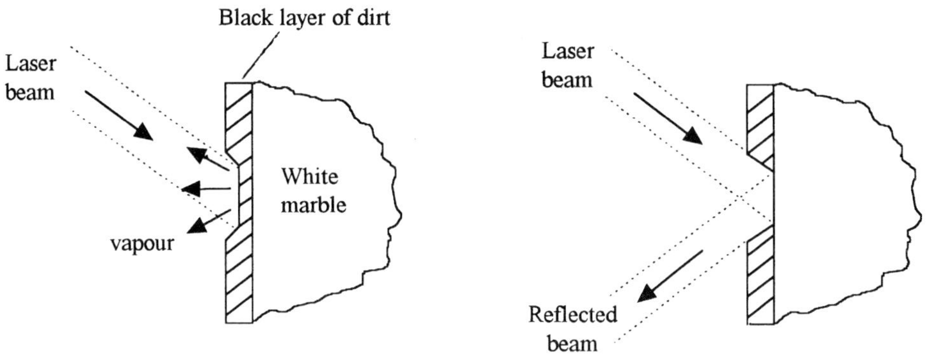

Absorption of laser beam

Initial interaction of long pulse radiation with a dark encrustation. Strong absorption of energy leads to vaporisation of material.

Reflection of laser beam

Final interaction of long pulse radiation with a dark encrustation. Once the encrustation has been removed further pulses are reflected from the weakly absorbing marble surface.

Figure 1.3. *Normal-mode cleaning (Asmus, 1973).*

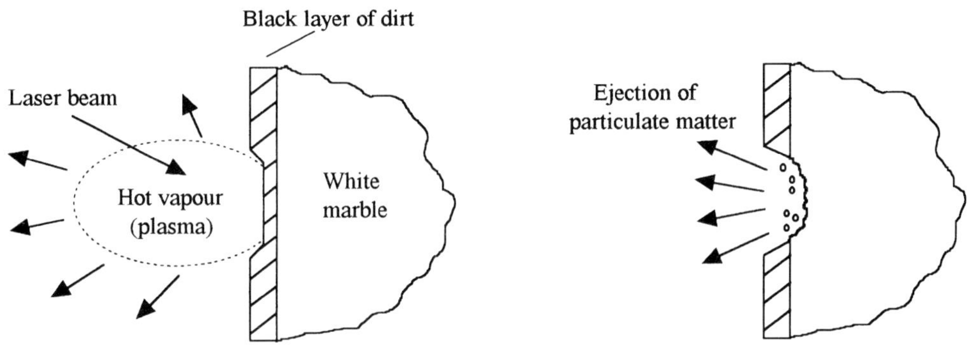

Beginning of laser pulse

Vaporisation of encrustation occurs early during the pulse, leading to formation of a plasma. The temperature and pressure of the plasma increase rapidly as the incoming laser radiation is absorbed and a microscopic compression is applied to the surface.

End of same laser pulse

As soon as the laser pulse finishes the plasma expands away from the surface. The surface relaxes and a thin layer of material is ejected.

Figure 1.4. *Removal of material by Q-switched laser radiation (Asmus, 1973).*

marble; whereas, repeated Q-switched pulses may continue to remove a small thickness of additional material (probably no more than a few microns)'. However, during the course of this work initial cleaning was carried out on a small marble gargoyle from the Loggetta del Campanile di San Marco using the laser in both modes, the most successful results being achieved in Q-switched mode. The cleaning was described as 'quite thorough without causing discernible damage'. Researchers have since suggested that there is more than one mechanism by which cleaning proceeds in Q-switched mode and that the dominant mechanism, and hence selectivity of the process, is dependent on the energy density (the energy per unit area falling on the surface) of the laser beam (see Chapter 3). Asmus also noted that the laser technique was useful 'in removing surface debris from a number of materials that for one reason or another had been virtually untreatable, previously'. These included: cleaning an especially friable calcium sulphate surface on marble, removal of light-coloured mineral deposits from a fourteenth-century Cyprian terracotta and removal of mineral deposits from a strip of lead on a twelfth-century sarcophagus. This work demonstrated the potential of the laser in a wide range of cleaning applications within conservation and also highlighted a number of areas which required further study:

1. the effectiveness of the technique on different types and colours of marble, and also on different materials;
2. the effectiveness of different wavelengths and pulse lengths in cleaning;
3. the effects of laser cleaning on the surface of impregnated marble;
4. the optimum energy densities for cleaning in each case.

In 1975 Asmus published papers in which he investigated the possible application of lasers to the conservation of stained glass (Asmus, 1975). Asmus suggested that the ability of the laser to deposit 'spatially and temporally localized energy in a material' could be used to treat deteriorating surface flashing on antique stained glass. He argued that if a sufficiently short pulse length was used then it would be possible to remove surface mineralization through vaporization without revitrifying the glass beneath and that at longer pulse lengths, which allow conduction of heat into the material, it would be possible to revitrify a mineralized layer of flashing to a specified depth. By solving the heat flow equation (Carslaw and Jaeger, 1959) for a one-dimensional model of a plane glass surface Asmus obtained an expression for the temperature rise induced by absorption of a single laser pulse as a function of the incident energy density. This was rearranged to give a relationship between incident laser energy density and surface, absorptivity for a particular temperature rise, assuming the density, thermal conductivity and specific heat capacity of the glass are known. Asmus claimed that if the surface absorptivity of the glass was measured then this relationship could be used to determine the laser energy density necessary to either remove or revitrify surface material. Unfortunately a thorough quantitative evaluation of this theory was not carried out. Initial tests were undertaken in which a ruby laser (pulse length 1 ms) was used to remove superficial layers from fragile samples of painted and enamelled glass. These exploratory experiments were considered to be a success and Asmus concluded that 'as there are no indications of copious surface damage, it may be possible safely to clean relatively large areas where variations in material properties would suggest variations in the optimum laser flux'. The ruby laser used in this work was not able to produce pulses long enough for surface vitrification so this area was not investigated.

In the same year, initial work was also undertaken on fresco samples dating from the thirteenth century (Asmus, Westlake and Newton, 1975). The aim of this work was to investigate the possibility of using pulsed laser radiation to remove lime coatings from frescos, in particular to uncover a Leonardo da Vinci mural (*The Battle of Anghiari*) in Florence, Italy. A similar analysis to that developed in the preceding paper was used for guidance during this study, with Asmus noting that although it was necessary to make many assumptions his solution had 'the virtue of simplicity' and that 'a more exact numerical computer solution might yield no better agreement with experiment as the thermal and optical parameters for most naturally occurring minerals are only approximately known and are quite poorly known for the heterogeneous and variable materials of interest here'. Assuming the lime and fresco materials to have the same thermal conductivity and diffusivity, Asmus concluded that if the lime covering had a sufficiently low vaporization temperature or high absorptivity compared with the fresco then differentiation and divestment would be possible. Asmus found that since the lime was white and had very low absorption at the ruby laser wavelength (690 nm), very high energy densities (17–20 J/cm^2) were required to remove the lime using normal-mode pulses (1 ms). Removal of the lime covering was accompanied by removal of the underlying pigment. In order to increase the absorptivity of the lime, with respect to the fresco, a saturated solution of copper sulphate was applied to the lime (the organic materials within the fresco prevented penetration beyond the lime). Removal of the lime was then achieved at a significantly lower energy density (7–9 J/cm^2) without removal of the pigment. Some gypsum grains of the fresco were loosened during the process but this problem was overcome by covering the surface with a sheet of rice paper moistened with water and then firing the laser through this layer. Asmus noted that 'generally, a few grains of lime would be left in contact with the surface after one or two pulses. However, as soon as the continuity and integrity of the lime layer were broken, it was a simple matter to simply brush away the residue' and Asmus suggested that it would be possible to uncover the Leonardo da Vinci fresco by removing all but the last few millimetres of lime by hand and then using a pulsed ruby laser to achieve the final divestment.

In 1976 research was undertaken by Asmus and co-workers investigating the surface morphology of laser-cleaned stone (Asmus, Seracini and Zetler, 1976). The absorptivities of a number of stone surfaces at the ruby laser wavelength were measured and, using the analysis developed in previous papers, the results used to predict (1) the energy density necessary to remove a black encrustation and (2) the energy density above which the clean stone surface would be damaged. Tests carried out on a Venetian marble sample, using a pulsed ruby laser (pulse length 1 ms), were in agreement with theory: vaporization of the black crust material occurred at a significantly lower energy density (6.8 J/cm^2) than vaporization of the marble (17 J/cm^2). Using Scanning Electron Microscopy (SEM), Asmus studied surface features on marble samples before and after laser cleaning and found that if cleaning was carried out close to the vaporization threshold of the black crust then there was no evidence of damage to the stone surface. He also noted that when cleaning was carried out close to the vaporization threshold for marble, microcrystalline damage was observed but that when compared to the effects of air-abrasive cleaning on friable marble 'even the microscopic surface modification by normal-pulse laser cleaning at the higher fluxes seems nominal'.

Another area in which initial work with a pulsed ruby laser met with some success was the cleaning of leather and vellum (Vitkus and Asmus, 1976). A normal-mode pulsed ruby laser was used to remove fungi from vellum. The authors noted that an energy density of 0.3–0.5 J/cm^2 'removed the fungi without discernible disruption of the skin' and upon inspection by SEM it became clear that laser cleaning disturbed the bulk and surface fibres substantially less than cleaning with an aqueous solution. The same laser was also successfully used to remove mildew from leather, although the maximum energy density below the onset of damage was found to be only 0.1–0.2 J/cm^2. The authors suggested that the laser technique 'offers the possibility of chemical-free mildew destruction' and might, therefore, avoid problems of 'discoloration, removal of salts beneficial to preservation, abrasion and chemical changes in general treatment procedures'. They also warned that a technique such as this which involves optically-induced heating would only work on those areas which are accessible to the laser radiation.

The research carried out by Asmus and co-workers between 1972 and 1976 led to the development of a prototype 'statue cleaner' (Asmus, 1976). A normal-mode ruby laser (pulse length 0.6–1.0 ms) was used to supply high energy pulses and a 'beam manipulator' comprising two rotary joints and two prisms was used to guide the laser beam. Selection of this type of laser was made on the grounds that with this combination of wavelength and pulse length the cleaning action is 'highly self-limiting and can be controlled so as to yield undetectable damage to marble grains'. Other laser systems were considered but were not selected either for practical reasons, such as the liquid dye laser which required frequent dye replacement in clean-room conditions, or simply because satisfactory cleaning was not possible, for example, carbon dioxide lasers, although highly efficient, emit radiation in the far infrared at a wavelength where selective material removal was not achieved. The prototype system was capable of producing a 4.5 J pulse at seven-second intervals and was successfully used to clean a small marble statue in approximately 14 hours.

However, cooling requirements and a separate cabinet, weighing 120 kg, incorporating the power supply and control unit for the laser meant that the prototype system was 'hardly suitable where portability was desired'. International funding allowed Asmus to overcome some of these problems by developing a system based on a normal-mode Nd:YAG laser (see Chapter 2; Asmus, 1976; Asmus, Lazzarini, Martini and Fasina, 1977): cooling requirements were simplified and the size of the power supply reduced so that it was possible to mount the laser head and beam manipulator on a tripod. Asmus found that the Nd:YAG laser gave very similar results to the ruby laser. The system was capable of producing one pulse per second and up to 10 J of energy in each pulse, i.e. fifteen times the coverage of the prototype system. The performance of the 'statue cleaner' was promising; removal of a thin encrustation from a fragile Greek marble capitol revealed a pleasing creamy-coloured patina and was achieved roughly four times faster than a similar area cleaned with a scalpel.

The work carried out by Asmus and co-workers between 1972 and 1978 demonstrated laser cleaning to be an extremely promising technique in conservation. The development of a technique capable of discriminating between a layer of dirt and the surface of an artefact, thereby allowing a skilled conservator to work on objects previously deemed 'beyond hope', was a very exciting prospect. Asmus had shown that the laser provided a significant improvement over other techniques in the cleaning of a wide range of materials. However, he warned that since the wavelength and pulse length determine the suitability of a certain type of laser for a particular job, and that a single laser which covers the entire parametric range did not exist, the laser was by no means the 'completely universal cleaning system' (Asmus, 1978). At that time the lasers most suitable for cleaning applications were relatively large, temperamental when used in a dusty conservation environment and very expensive. Although practical projects were undertaken using pulsed Nd:YAG systems the cleaning process was slow and difficulty was experienced in obtaining a uniform result (Asmus, 1986). Other techniques, such as air-abrasive cleaning, although not as selective as the laser technique, provided the conservator with a more practical and affordable solution. Asmus concluded that 'surface divestment with light, in all probability, will become exceedingly important when more versatile laser types become commercially available. In the meantime there are a number of specialized problems where current laser technology is able to complement the customary approaches'.

The initial work undertaken by Asmus and co-workers during the 1970s clearly demonstrated the potential of laser-based cleaning techniques, and the theoretical model proposed at this time, although making many assumptions, provided a basic understanding of the processes involved. Asmus acknowledged that the interaction of a high power laser beam with a surface of variable and sometimes unknown composition is indeed a complex process, with the actual mechanism responsible for material removal depending on beam parameters such as wavelength, pulse length and energy density. By the late 1970s much useful research into the laser cleaning of artworks had been carried out but questions still remained unanswered, particularly in the areas of pulse length dependence, wavelength dependence and practical cleaning: would a mobile, reliable and affordable system capable of cleaning at least as quickly as other techniques ever be available?

Research undertaken since 1980

In 1981, a report from the Canadian Conservation Institute (Carlyle, 1981) described initial work in which excimer lasers (emitting radiation in the ultraviolet part of the spectrum) were used to remove dirt and varnish from oil paintings. The results were described as encouraging and the author felt that the potential impact of the technique on the conservation community warranted further study as soon as possible. This work reinforced the view of Asmus that laser cleaning was an extremely promising technique but little research was actually carried out in this area during the 1980s. Although laser cleaning was acknowledged to have many advantages over other methods it was a relatively slow, costly and in many cases completely impractical process to employ during that time. It was not until the late 1980s, when concern over possible damage caused to objects by established cleaning methods was increasing, and when laser technology had developed to a sufficient extent, that interest was revived and research undertaken, mainly in the United Kingdom, France and Italy, to look once more at the feasibility of using the laser as a conservation tool.

In 1989 research was carried out at the Laboratoire de Recherche des Monuments Historiques (Orial and Gaufillet, 1989) using a Nd:YAG laser in both normal and Q-switched modes to clean surfaces of marble, limestone, silver and stained glass. Using the laser in normal-mode (0.4–12 ms) pollution encrustations were removed from marble and limestone without any noticeable alteration to either the surface morphology or the chemical composition of the stone. In contrast, SEM revealed 'micro-craters' in areas cleaned by air-abrasives. Tests on silver (tarnish not removed) and glass (damaged by thermal shock) were unsuccessful. To investigate the possibility of working on-site the laser was transported to la Collegiale de Thann in the back of a small van where tests were carried out on a badly polluted piece of nineteenth-century sculpture. A combination of laser radiation and nebulized water successfully removed the black encrustation in a manner which suggested that development of a system specifically adapted for work in-situ would be of great benefit to the conservation world: during the course of this work it became evident that by employing this technique work could be carried out in an extremely precise and controlled manner and relatively quickly. Tests carried out using the laser in Q-switched mode (pulse length 12 ns) were even more encouraging. Black pollution layers were successfully removed from friable pieces of limestone (without nebulized water), hard mineral deposits were removed from the surface of ancient glass and corrosion layers were removed from a lead plaque without any apparent alteration to the basic metal. This work concluded that the cleaning of historic monuments is best achieved using the laser operating in Q-switched mode, since the short pulse length ensures there is no or very little temperature rise in the underlying surface and therefore no risk of burning. The laser used in this study was capable of emitting a maximum pulse energy of 900 mJ at a repetition rate of up to 30 pulses per second. The cleaning rates (100 cm^2 in 1.3 minutes for a lightly soiled limestone surface and 5 minutes for a heavily encrusted surface) were faster than those achieved by Asmus and co-workers in 1976. The authors suggested that such cleaning rates were sufficient for sculpted surfaces and that development of a commercial mobile laser cleaning system should be investigated.

In the United Kingdom initial work concentrated on the suitability of different types of laser for the removal of black encrustations from limestone sculpture (see Figure 1.5) (Cooper, Emmony and Larson, 1992a; 1992b). Tests were carried out using four different types of pulsed laser, with wavelengths ranging from the ultraviolet to the far infrared part of the spectrum: excimer (wavelength 248 nm, pulse length 30 ns), dye (590 nm, 20 μs), Nd:YAG (1.06 μm, 10 ns/100 μs) and carbon dioxide (10.6 μm, 100 ns). The work showed that it was possible to remove the dirt layers at each of the wavelengths, but closer inspection of the cleaned surfaces through a binocular microscope revealed that the most selective cleaning was achieved with the Nd:YAG laser. Furthermore, it was clear that cleaning was much more effective when the laser was operated in Q-switched mode (pulse length 10 ns). Tests carried out with the short pulses on a friable piece of stone left the surface with a clean, uniform patinated appearance. In contrast, tests with the longer normal-mode pulses left a greyish part-cleaned surface which appeared to have suffered slight burning. The

Figure 1.5. *Limestone carving (70 cm tall) of a grotesque from Lincoln Cathedral showing a typical thick black encrustation resulting from atmospheric pollution. Reproduced by kind permission of the Dean and Chapter of Lincoln.*

authors, in agreement with Orial and Gaufillet, concluded that the Q-switched Nd:YAG laser was the most suitable type of laser for the cleaning of limestone sculpture; at 1.06 μm a typical pollution encrustation absorbs significantly more of the incident laser radiation than the underlying limestone surface, so that selective removal is possible, and the short pulse length ensures that thermal conduction into the limestone surface is negligible.

Research carried out during the following year further demonstrated the selective nature of laser cleaning at 1.06 μm (Cooper, Emmony and Larson, 1993a; 1993b; 1995). Cooper and co-workers measured the minimum energy densities required to remove material from a black crust and from a limestone surface and found that a range existed within which the process became 'self-limiting', i.e. once the black crust had been removed further pulses were unable to remove material from the limestone. Cooper and co-workers also found that the audible 'snapping sound' associated with rapid ejection of material from a surface disappeared once the limestone substrate had been revealed, provided cleaning was carried out within the 'self-limiting' regime. The authors suggested that the loudness of this sound could be used by the conservator as a means of ensuring that cleaning is carried out within this regime, for example if, after the black crust has been removed, each pulse is still accompanied by a loud crack then the energy density is too high and should be reduced immediately to avoid damage to the limestone.

The encouraging nature of the results obtained during this research led to the development of a prototype laser cleaning tool (see Figure 1.6) (Cooper, 1994). The Nd:YAG laser head was mounted on a trolley and connected to a separate power supply by a flexible tube which housed the power leads and cooling water pipes. A six-jointed articulated arm delivery system was used to guide the beam which emerged from a pen-like handpiece. The system was powered from a 13A/240V mains power supply and provided laser radiation at a wavelength of 1.06 μm in pulses of up to 300 mJ energy, of 5–10 ns duration at a repetition rate variable between 0–10 pulses per second. Systems such as these provided the conservator with a practical means of using laser radiation to clean sculpture; cleaning could now be undertaken in a conservation studio or even on-site rather than in the confines of a scientific research laboratory. In addition, rates of cleaning were comparable with, and in many cases faster than those of, more conventional techniques. The realization by conservators that laser cleaning had reached a stage where it could now be considered a practical and affordable technique (Cooper and Loton, 1994; Teppo and Calcagno, 1995) created a tremendous amount of interest in the area and commercial laser cleaning systems have since been developed in the United Kingdom, France and Italy specifically for use in conservation (see Figure 1.7).

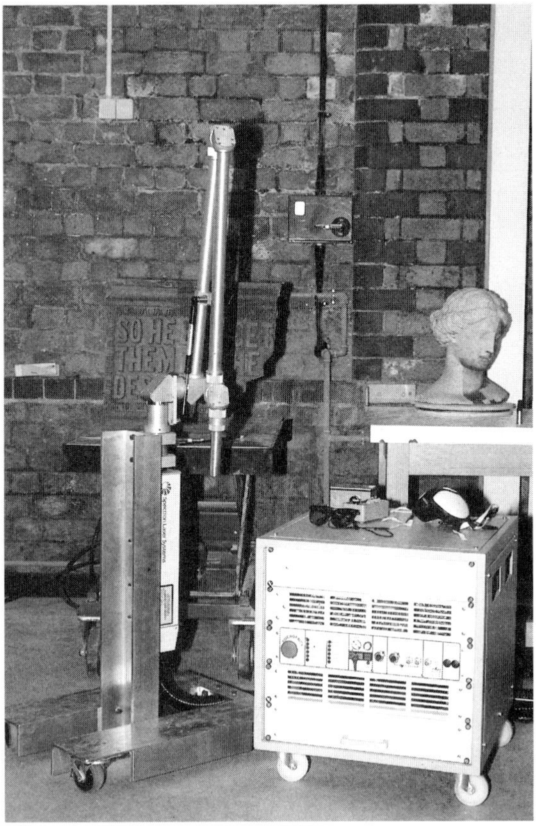

Figure 1.6. *A prototype Nd:YAG laser-cleaning system developed in the United Kingdom for the cleaning of limestone sculpture.*

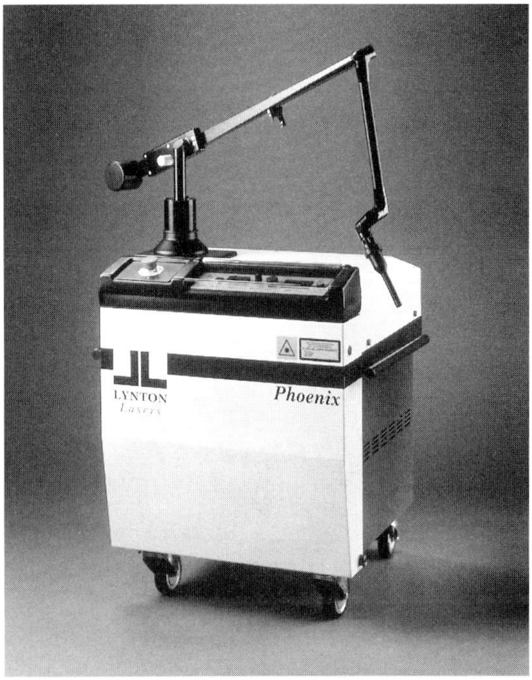

Figure 1.7. *A commercial Nd:YAG laser-cleaning system developed specifically for use in conservation.*

Between 1993 and 1996 research into the use of lasers for the treatment of a wide range of materials encountered in conservation has been undertaken. Initial work tended to concentrate on sculpture but interest has since developed in areas as diverse as paintings, paper, silk and stained glass.

Research at near infrared wavelengths

Lasers and stone

In 1993 research was published in which the effects of laser cleaning (Q-switched Nd:YAG, wavelength 1.06 μm, pulse length 8–25 ns) and micro air-abrasive cleaning (alumina powder, 6 and 27 μm diameter) on polluted marble and limestone were compared (Verges-Belmin, Pichot and Orial, 1993). Tests were carried out on thirteenth-century limestone from Amiens cathedral and on a nineteenth-century Carrara marble statue from L'Ecole des Beaux-Arts in Paris. In both cases the surface of the stone had reacted with atmospheric gases to form a gypsum film, or patina, within which the surface detail of the carving had been preserved. A black encrustation had formed on top of this layer. Petrographic techniques were employed to assess the level at which the cleaning techniques stopped. The results showed that laser cleaning was, in both instances, a much more selective process than micro air-abrasive cleaning. On the laser-cleaned samples the cleaning stopped as soon as the encrustation had been removed and the patina revealed, thereby preserving the surface detail. With the abrasive-cleaned samples, however, the cleaning continued until the stream of abrasive was directed elsewhere. This resulted in significant damage to the patina and loss of surface detail.

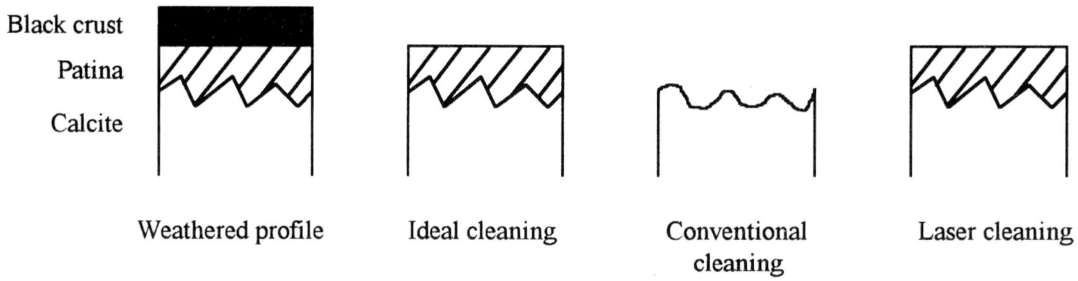

Figure 1.8. *Preservation of patina on marble by laser-cleaning (Verges-Belmin, et al., 1993; Cooper and Larson, 1996a; b).*

Similar results were obtained by Cooper and Larson during cleaning of a heavily polluted nineteenth-century Carrara marble statue of Lord Huskisson (Cooper and Larson, 1996; 1997). Cleaning was again carried out with a Q-switched Nd:YAG laser operating at 1.06 μm. Microscopic examination of cross-sections through samples from the statue revealed that the patina had been preserved (see Chapter 5). Areas of the statue which had previously been steam-cleaned, however, had been stripped of the patina. These results are shown schematically in Figure 1.8. Cooper and Larson also described how other forms of patina, from the natural ageing of ancient Greek sculpture to artificially aged eighteenth-century restorations, have been preserved by carefully controlling the energy density of the laser beam on the sculpture surface during cleaning. The authors argued that at relatively low energy densities dirt layers are removed by highly selective mechanisms whereas at higher energy densities removal occurs via less controllable mechanisms such as the plasma-dominated process proposed by Asmus and co-workers in 1973. In order to preserve patina on sculpture, cleaning should therefore be undertaken at the lowest possible energy density that will remove dirt layers and allow completion of the work in the time available.

Lasers and terracotta

Research has also shown that extremely precise and controlled cleaning of terracotta surfaces can be achieved by careful use of a Q-switched Nd:YAG laser (Larson and Cooper, 1996). The research showed that, provided cleaning is carried out below a certain threshold energy density, it is possible to (1) clean terracotta surfaces in sound condition without damaging the protective fireskin and (2) clean terracotta surfaces in poor condition without damaging the surface further. The authors noticed that if cleaning was undertaken at slightly too high an energy density then pits (less than 0.5 mm in diameter) were observed in the cleaned surface from where small amounts of terracotta had been removed. At much higher energy densities larger amounts of material were removed and adjacent areas underwent a colour change from red to silvery-grey. This was attributed to changes in the oxidation state of the iron in the surface as a result of heating caused by the excess energy deposited by the laser beam.

The work also demonstrated the importance of pulse length in laser cleaning. While it was possible to remove dirt layers efficiently without damaging the underlying terracotta using Q-switched pulses, Larson and Cooper found that attempted removal of the same layers by much longer normal-mode pulses (100 μs duration) led to the formation of a cracked glassy layer on the surface of the terracotta which, upon inspection by SEM, was found to have a crazed appearance. The authors argued that if energy is deposited into a surface layer over a relatively long time (as is the case with normal-mode pulses) then conduction of heat into the substrate becomes a very significant factor in the process. In this instance conduction of heat from the dirt layers into the terracotta appeared to lead to revitrification of the terracotta surface.

Lasers and textiles

The interaction of laser radiation with silk has also been studied (Oger and Polonowski, 1994). In 1994 research was described in which a Nd:YAG laser (pulse length not known) was used to remove dirt crusts and stains from a silk model. The work met with mixed success. On one area of the model the following observation was made (translated into English): '[According to] the conditions of use one can obtain results varying from perfect cleaning without any modification of the cohesion of the thread to the destruction or annihilation of the layer of the model'. Inspection of silk by SEM from other areas of the model cleaned by laser radiation confirmed that there was 'no degradation of the silk fibres and in particular no de-fibrillation of the silk fibre'. The results were considered promising but the authors stressed the need for examining the effect of laser radiation on old and fragile organic material in a laboratory before proper cleaning is undertaken since damage, such as loss of cohesion of pieces of silk, might only be visible through a binocular microscope. The authors also suggested that 'considering the damage possible at excessive exposures' use of the laser should be restricted to 'qualified conservators who have a good knowledge of both the materials and the machine'.

The results of initial tests on other materials, including old textiles, tapestries and stained fabric, were also briefly described. The surfaces of thick textiles and tapestries were successfully cleaned, thereby enhancing their visual appearance. Dust was not removed from beneath the surface, however. At high intensities the presence of glue stains led to burning of the wool and chemically bound stains, such as ink, rust and micro-organic stains, were not removed by radiation at a wavelength of 1.06 μm. The authors felt that the effect of laser radiation on glues and some dyes required further research.

Research at visible and ultraviolet wavelengths

Much work has gone into the development of the Nd:YAG laser which, over the course of the last twenty-five years, has become a reliable, compact and relatively efficient type of laser. These practical attributes, combined with the advantages of radiation at 1.06 μm in terms of strong absorption by many dirt layers and weak absorption by many artefacts have meant that most research in laser cleaning has been concerned with the effects of laser radiation at 1.06 μm. However, work has also been undertaken at other, mainly visible and ultraviolet, wavelengths. This work has concentrated on artefacts, such as paper, stained glass and paintings, where certain components are either strongly absorbing at 1.06 μm or susceptible to heating effects caused by infrared radiation and where extremely precise layer-by-layer removal of dirt is desirable.

Lasers and paper

In 1994 research was published in which the frequency-doubled output from a Nd:YAG laser (532 nm, green) and the output from a dye laser (566 nm, yellow) were successfully used to remove certain types of fungus-induced stains from paper (Szczepanowska and Moomaw, 1994). Tests were carried out on four different colours of staining: dense black, pink, light green and brownish-grey on a medium weight wove paper. The results were examined under a binocular microscope and using SEM. In each case the 'intensity, rate and number of pulses were varied to determine the optimal conditions for stain removal and to avoid damage to the paper'. Radiation at 532 nm was found to be very effective in removing two of the stains: black and light green, and radiation from the lower power dye laser (566 nm), although less effective in removing stains, was able to remove fungal bodies from the pink stain. In each case neither damage nor rearrangement of the paper fibres was observed. The authors described the degree of removal of fungus and lack of damage to the paper as 'truly remarkable' and suggested that 'removal is far superior to any mechanical or solvent methods that have been tried to date'. Tests were also carried out on an area of heavy black printer's ink to determine possible damage to script. The authors found that 6–10 times more pulses were required to remove a noticeable amount of ink than were needed to remove the fungal stains. The research concluded that 'laser stain removal has significant advantages for removing stains from artwork on paper' and by selecting an appropriate wavelength for a particular stain and optimizing the energy density used it should be possible to remove a wide range of disfiguring stains.

Lasers and polychrome surfaces

Research into laser cleaning of stained glass and paintings has been carried out using excimer lasers which provide a source of ultraviolet radiation. At these shorter wavelengths cleaning occurs by directly breaking chemical bonds within the dirt and involves less heating than at infrared wavelengths: a process known as 'cold' ablation. Dirt layers can, therefore, be removed a single micron thickness at a time, thereby providing adequate warning for the process to be stopped before the paint or glass surfaces are exposed to laser radiation. In many cases parts of a painting or stained glass panel will absorb strongly at the chosen wavelength so that a 'self-limiting' regime does not exist and some form of monitoring and process control is necessary to avoid damaging the artefact.

In 1994 research was carried out to investigate the removal of 10 μm thick encrustations from medieval stained glass using krypton fluoride (KrF) excimer laser radiation at a wavelength of 248 μm (Dickmann, Olaineck and Bachmann, 1994). A threshold energy density for crust removal of 0.6 J/cm^2 and a removal rate of 0.7 μm/pulse at 3 J/cm^2 were measured. The work showed that by altering the process parameters, such as energy density and number of pulses, to take account of the inhomogeneous nature of the crust and variations in its thickness, laser assisted cleaning of the glass surface was possible. The authors stated that 'for practical applications a closed loop process control is necessary to avoid any damage of the glass substrate' and predicted that since the technique is relatively slow applications would probably be restricted to 'the field of extremely valuable cultural assets'. In a separate study (Leissner and Fuchs, 1995) laser radiation at 248 nm was used to remove organic conservation coatings and inorganic corrosion crusts from glass surfaces. The research found that the threshold energy density for organic coating removal was sufficiently low to avoid any damage to the underlying gel layer and bulk glass. The energy density required to remove corrosion crusts, however, was also sufficient to remove the gel layer and occasionally damage the bulk glass. The authors suggested that Fourier Transform Infrared Spectroscopy (FTIR) is an appropriate characterization technique to monitor the removal of different materials from the glass surface and hence prevent any damage.

Ultraviolet laser radiation has been successfully used in the conservation of paintings (Morgan, 1993; Fotakis, Hontzopoulos, Zergiotti *et al.*, 1995). Fotakis and co-workers have identified three main areas where a KrF excimer laser (248 nm, pulse length 15 ns) can be used:

1. cleaning of surface layers of varnish or contaminated regions of a painting;
2. cleaning of a painting's support material;
3. removal of overpainting.

A typical varnish or paint stroke applies approximately 15–20 μm of material to the surface. By controlling the energy density on the surface, Fotakis and co-workers were able to control the depth of material removed: 0.5–1.0 μm/pulse. The authors have shown that, when combined with on-line monitoring techniques, this provides adequate control for the removal of surface layers without damaging the underlying pigment. A seventeenth-century icon has been successfully cleaned and a nineteenth-century overpainting successfully removed from a fourteenth-century icon. Fotakis and co-workers predicted 'the ultimate development of safe, portable workstations incorporating both cleaning and diagnostic modules to allow works of art to be restored on site' but warned that extensive work still remained to be done, including 'the scientific categorization of pigment, varnish and substrate data so the correct laser conservation strategies can be selected for the task at hand'. The cleaning of paintings is described in more detail in Chapter 6.

Conclusion

In October 1995 the first international conference on Lasers in the Conservation of Artworks, LACONA, was held in Crete, bringing together scientists and conservators to discuss the major advances made during the 1990s and possible developments in the future. Laser cleaning has come a long way since the early days of Asmus during the 1970s; commercial laser cleaning systems are now available and the technique has become established (particularly in the area of sculpture conservation) in a number of conservation studios spread across Europe. Research is currently being carried out in many areas. At the Crete conference papers were presented in fields as diverse as; Nd:YAG cleaning of marble sculpture; excimer cleaning of stained glass; laser cleaning of wall paintings; excimer cleaning of antique parchment; removal of mud, fungi and stains from paper by excimer laser; laser cleaning of metals at infrared wavelengths; excimer cleaning of paintings and even the cleaning of ancient skulls by excimer lasers. Much research remains to be done and although Nd:YAG lasers are becoming an established tool, laser systems operating in the ultraviolet have yet to reach the stage where they can be routinely used in a conservation studio. As the technology develops further and prices come down this is surely just a matter of time.

References

Ashurst, J. and Dimes, F. (1990) *Conservation of Building and Decorative Stone, Volume 2*, Butterworth-Heinemann

Asmus, J.F. (1975) Use of lasers in the conservation of stained glass. *Conservation in Archaeology and the Applied Arts*, June, 139–42

Asmus, J.F. (1976) The development of a laser statue cleaner. *Proceedings of the 2nd International Symposium on the Deterioration of Building Stones*, Athens, 137–41

Asmus, J.F. (1978) Light cleaning: laser technology for surface preparation in the arts. *Technology and Conservation*, **3** (3), 14–18

Asmus, J.F. (1986) More light for art conservation. *IEEE Circuits and Devices Magazine*, **2** (2), March, 6–15

Asmus, J.F., Guattari, G., Lazzarini, L., *et al.* (1973) Holography in the conservation of statuary. *Studies in Conservation*, **18**, 49–63

Asmus, J.F., Lazzarini, L., Martini, A. and Fasina, V. (1977) Performance of the Venice statue cleaner. *Preprints of the Fifth Annual Meeting of the American Institute for Conservation of Historic and Artistic Works*, Boston, Massachusetts, 30 May–2 June, 5–11

Asmus, J.F., Murphy, C.G. and Munk, W.H. (1973) Studies on the interaction of laser radiation with art artifacts. *Proceedings of SPIE*, **41**, 19–27

Asmus, J.F., Seracini, M. and Zetler, M.J. (1976) Surface morphology of laser-cleaned stone. *Lithoclastia*, **1**, 23–46

Asmus, J.F., Westlake, D.L. and Newton, H.T. (1975) Laser technique for the divestment of a lost Leonardo da Vinci mural. *Journal of Vacuum Science and Technology*, **12** (6), 1352–55

Bluck, B. and Porter, J. (1991a) Sandstone buildings and cleaning problems. *Stone Industries*, **26** (2), March, 21–27

Bluck, B. and Porter, J. (1991b) Aims and methods of sandstone cleaning. *Stone Industries*, **26** (3) April, 21–28

Carlyle, L. (1981) Laser interactions with paintings: results and proposals for further study. Unpublished report at the Canadian Conservation Institute, Ottawa, 16 November 1981

Carslaw, H.S. and Jaegar, J.C. (1959) *Conduction of Heat in Solids*, 2nd edition, Oxford University Press

Cooper, M.I. (1994) *Laser cleaning of stone sculpture*. PhD Thesis, Loughborough University, 188–208

Cooper, M.I., Emmony, D.C. and Larson, J.H. (1992a) The use of laser energy to clean polluted stone sculpture. *Journal of Photographic Science*, **40**, 55–57

Cooper, M.I., Emmony, D.C. and Larson, J.H. (1992b) A comparative study of the laser cleaning of limestone. *Proceedings of the 7th International Congress on Deterioration and Conservation of Stone* (J. Delgado Rodrigues, F. Henriques and F. Telmo Jeremias, eds), Lisbon, June 1992, 1307–15

Cooper, M.I., Emmony, D.C. and Larson, J.H. (1993a) Laser cleaning of limestone sculpture. In *Conservation Science in the UK* (N.H. Tennent, ed.). Preprints of meeting held in Glasgow, May 1993, 29–32, James & James

Cooper, M.I., Emmony, D.C. and Larson, J.H. (1993b) The evaluation of laser cleaning of stone sculpture. In *Structural Repair and Maintenance of Historical Buildings III* (C. Brebbia and R. Frewer, eds), 259–66, Computational Mechanics Publications

Cooper, M.I., Emmony, D.C. and Larson, J.H. (1995) Characterization of laser cleaning of limestone. *Optics and Laser Technology*, **27** (1), 69–73

Cooper, M.I. and Larson, J.H. (1996) The use of laser cleaning to preserve patina on marble sculpture. *The Conservator*, **20**, 28–35

Cooper, M.I. and Larson, J.H. (1997) Laser cleaning of marble sculpture. *Restauratorenblatter*, Proceedings of LACONA I (Crete, October 1995), International Institute for Conservation of Historic and Artistic Works, Verlag Mayer, Vienna

Cooper, M.I. and Loton, A. (1994) Light years ahead? *Natural Stone Specialist*, December, 24–30

Dickmann, K., Olaineck, C. and Bachmann, F. (1994) Reinigung von Glasoberflachen an kulturhistorischen Gutern mit Excimer-laser-strahlung. *Laser Magazin*, June, 10–13

Einstein, A. (1917) Zur Quantentheorie der Strahlung. *Physische Zeitschrift.*, **18**, 121

Engelsberg, A.C. (1995) Laser-assisted cleaning proves promising. *Precision Cleaning*, May

Fotakis, C., Hontzopoulos, E., Zergiotti, I., *et al.* (1995) Laser applications in painting conservation. *Proceedings of the AIC Conference*, June

Gagliano, F.P. and Zaleckas, V.J. (1972) Laser processing. In *Lasers in Industry* (S.S. Charschan, ed.) pp. 190–295, Van Nostrand Reinhold Company

Goldman, L. (1971) *Lasers in Medicine*, Gordon & Breach Science Publishers Inc

Imen, K., Lee, S.J. and Allen, S.D. (1991) Laser-assisted micron scale particle removal. *Applied Physics Letters*, **58** (2), 203–205

Jackson, M.R. (1994) Lace shows how lasers can cut flexible materials. *Opto and Laser Europe*, **15**, November, 21–24

Lal Gauri, K. and Holdren, G.C. (1981) Pollutant effects on stone monuments. *Environmental Science and Technology*, **15** (4), 386–90

Larson, J.H. and Cooper, M.I. (1996) The use of laser energy for cleaning architectural terracotta decoration. *Architectural Ceramics; Their History, Manufacture and Conservation*, Joint symposium of English Heritage and UKIC, September 1994, 92–100, James & James

Lazzarini, L. and Asmus, J.F. (1973) The application of laser radiation to the cleaning of statuary. *Bulletin of the AIC*, **13** (2), 39–49

Leissner, J. and Fuchs, D.R. (1995) Examination of excimer-laser treatments as a cleaning method for historical stained glass windows. *Fundamentals of Glass Science and Technology, 3rd ESG Conference*, 332–9

Maiman, T.H. (1960) Stimulated optical radiation in ruby masers. *Nature*, **187**, 493

Morgan, N. (1993) Excimer lasers restore 14th century icons. *Opto and Laser Europe*, 7 September, 36–37

Oger, B. and Polonovski, M. (1994) The feasibility of cleaning silk in topographical models by laser. *Preprints of the conference 'La Conservation des Textiles Anciens', Journées d'Etudes de la SFIIC*, Angers, October

Orial, G. and Gaufillet, J.-P. (1989) Nettoyage des monuments historiques par désincrustation photonique des salissures. *Technologie Industrielle Conservation Restauration du Patrimoine Culturel*, Colloque AFTPV/SFIIC, Nice 19–22 September, 118–25

Schweizer, G. (1995) CO_2 lasers reduce waste from aircraft paint removal. *Opto and Laser Europe*, **18**, March, 42–43

Skoulikidis, T. and Papakonstantinou-Ziotis, P. (1981) Mechanism of sulphation by atmospheric SO_2 of the limestones and marbles of the ancient monuments and statues. *British Corrosion Journal*, **16** (2), 63–69

Szczepanowska, H.M. and Moomaw, W.R. (1994) Laser stain removal of fungus-induced stains from paper. *Journal of the American Institute of Conservation*, **33**, 25–32

Teppo, E. and Calcagno, G. (1995) Restoration with lasers halts decay of ancient artefacts. *Laser Focus World*, June, 55–59

Verges-Belmin, V., Pichot, C. and Orial, G. (1993) Elimination des croûtes noires sur marbre et craie: a quel niveau arrêter le nettoyage? *Proceedings of the International Congress on the Conservation of Stone and Other Materials*, Paris, 29 June–1 July 1993

Verges-Belmin, V. (1994) Pseudomorphism of gypsum after calcite, a new textural feature accounting for the marble sulphation mechanism. *Atmospheric Environment*, **28** (2), 295–303

Vitkus, J.R. and Asmus, J.F. (1976) Treatment of leather and vellum with transient heating. *Preprints of 4th Annual Meeting of American Institute for Conservation of Historic and Artistic Works*, 111–17

2

Basic principles of a laser

Martin Cooper

The laser is a source of 'pure' concentrated light. The term LASER is an acronym for Light Amplification by Stimulated Emission of Radiation. Here the word light is used to describe that part of the electromagnetic spectrum which extends from the ultraviolet through the visible (which can be detected by our eyes) to the infrared; radiation refers to electromagnetic radiation, of which light is one example, and amplification by stimulated emission refers to the process by which this special form of light is generated: a laser amplifies light by a phenomenon known as stimulated emission. Since the first demonstration of laser action in 1960 development of lasers has been extremely rapid. There are now many different types: ultraviolet, visible and infrared lasers; low and high power lasers; short pulse, long pulse and continuous wave lasers, and an equally wide range of applications, from industry to medicine. In some applications, such as the cutting of steel by laser, power is important, while in applications such as surgery selectivity and precision are critical. What these applications have in common is the capability of the laser to control exactly where, when and how much of this 'pure' concentrated light falls onto the surface being treated.

Energy, light and the electromagnetic spectrum

Appreciation of the unique properties of laser radiation is helped by a basic understanding of the nature of light. This section aims to provide the reader with that level of understanding. Light is a form of *energy*. Energy can be thought of as 'the ability to do work', i.e. some kind of movement against an opposing force. For example, a moving car, which has a form of energy known as kinetic energy, is working to overcome forces of friction (between its tyres and the road surface) and air resistance. There are many different forms of energy: heat or thermal energy is associated with the motion of molecules within a substance, e.g. molecules in a hot solid are vibrating faster than those in a cold solid; potential energy is stored energy, e.g. a stationary car at the top of a hill has potential energy which is released as the car rolls down the hill; chemical energy is stored in food, e.g. when food is eaten its stored chemical energy is converted by our bodies into heat and kinetic energy. Other forms of energy include electricity (movement of electrons), sound (vibration of air molecules) and light. Energy cannot be created or destroyed, but it can be converted from one form to another. For example, a hydroelectric power station uses the potential energy of water in a reservoir above the station to generate electricity: as the water flows down the mountainside its potential energy is converted into kinetic energy. At the foot of the mountain this kinetic energy is used to drive a turbine in the power station. This generates electricity which can be distributed throughout the region and used to power

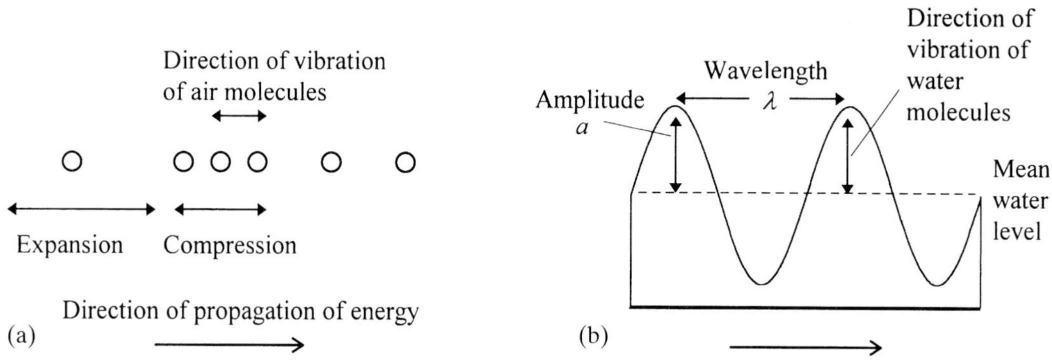

Figure 2.1. *Longitudinal and transverse wave motion.*

electrical appliances or converted into light and heat for thousands of homes.

Energy is transferred from one point to another, without any particles moving between the two points, by *waves*. For example, if a stone is dropped into a pond, waves spread out from the splash and eventually water at the edge of the pond moves as it receives some of the energy. Waves are everywhere. Some we can see or hear but many our senses are unable to detect. On a submicroscopic scale electrons which move in fixed orbits around the nucleus of an atom can be stimulated to emit X-rays, light and heat waves. At the opposite end of the scale earthquakes generate waves as a result of the sudden movement of land masses, and waves in the sea are produced by wind blowing across the surface.

There are two kinds of waves: longitudinal and transverse. In a longitudinal wave particles vibrate in a direction parallel to the direction in which the energy moves (referred to as the direction of propagation). Sound waves, which comprise a series of compressions and expansions of air molecules, are an example of longitudinal wave motion; energy is transferred parallel to the direction of vibration of the air molecules (see Figure 2.1a). In a transverse wave particles vibrate in a direction perpendicular to the direction in which the energy propagates. Water waves are an example of transverse wave motion; as the water molecules move up and down energy is transferred across the surface of the water (see Figure 2.1b). A cross-section through a water wave frozen in time would appear as in Figure 2.1b, made up of a series of crests (high parts of the wave) and troughs (low parts of the wave). The wave has certain characteristics: its amplitude, a, is the height of a crest above the mean water level, its wavelength, λ, is defined as the distance between two successive crests (or troughs) and its frequency, f, is the number of waves passing a fixed point in a single second. Light waves are another example of transverse wave motion.

Any wave motion is generated by some form of vibrating source (Jenkins and White, 1981) and the frequency of the waves will be equal to the frequency of that source. A relationship exists between the frequency, wavelength and velocity, v, of a wave: for a regular wave, its velocity is equal to the product of its wavelength and frequency ($v = \lambda f$). Since the frequency of the wave is determined by the vibrating source the wavelength in a particular medium depends on the velocity in that medium and can be determined by dividing the velocity by the frequency ($\lambda = v/f$). So, for a particular medium, if the frequency of the vibrating source increases then the wavelength of the waves generated decreases. This is shown schematically in Figure 2.2. Imagine a vibrating piston in a water tank. At low frequency only a few long water waves will be generated. At high frequency there will be many more water waves within the tank but with a shorter wavelength. The velocity of the waves will be unchanged.

Waves exhibit the properties of reflection, refraction, diffraction and interference. When a wave meets a boundary between two media

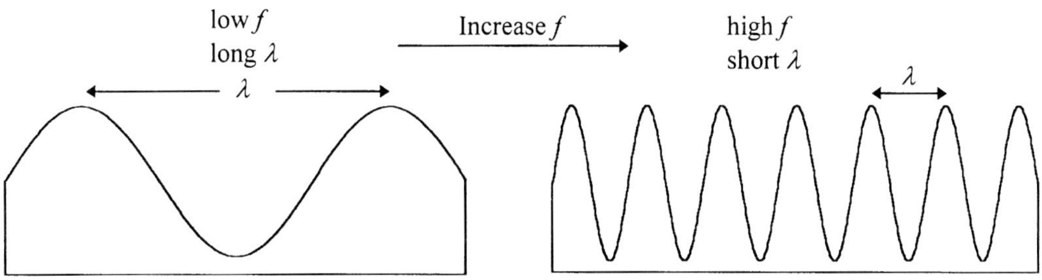

Figure 2.2. *Effect of increasing frequency, f, on the wavelength, λ, of water waves.*

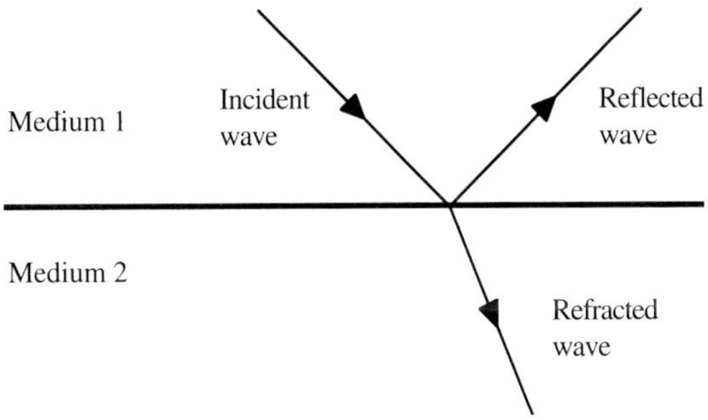

Figure 2.3. *Reflection and refraction of a wave at a boundary between two media.*

part of that wave will be reflected back into the first medium (see Figure 2.3): an example of this is the reflection of light by a mirror. The remainder of the beam is refracted, or bent in its path, as it enters the second medium (see Figure 2.3). A prism can be used to split white light into its constituent colours since each colour is bent, or refracted, to a slightly different extent. Diffraction is the spreading out of a wave after passing through a gap. The clearest diffraction occurs when the width of the gap is equal to the wavelength. Sound waves are diffracted as they travel through a doorway

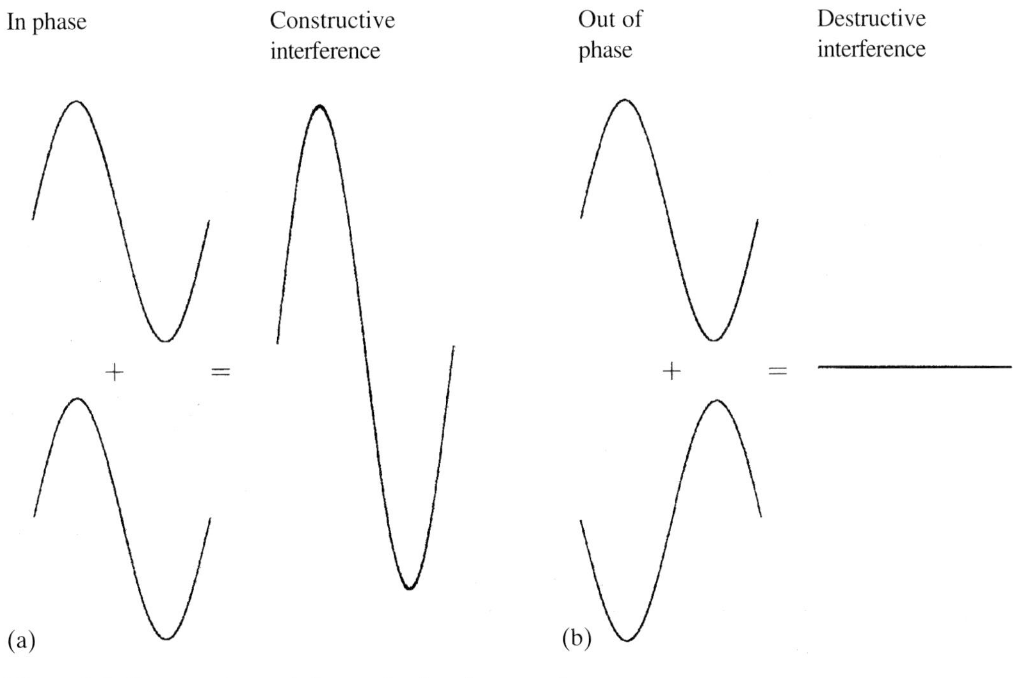

Figure 2.4. *Constructive and destructive interference of waves.*

making it possible to hear a person talking in a room even if you are standing in a corridor, out of sight. When two waves arrive at the same place at the same time they interfere. At places where the two waves arrive in step so that the crest of one wave overlaps the crest of the other wave, the waves interfere constructively and add together to form a bigger wave (see Figure 2.4a). In this case the two waves are said to be *in phase*. At places where the two waves arrive out of step so that the crest of one wave coincides with the trough of the other wave, the waves interfere destructively and cancel each other out (see Figure 2.4b). In this case the two waves are said to be *out of phase*.

To summarize: there are many different forms of energy, of which light is just one example; laser radiation is a unique form of light. Waves transfer energy from one point to another without particle movement between the two points. The velocity of the wave depends on the medium through which the wave is travelling. Waves are reflected and refracted at boundaries, diffracted through openings and when two or more waves meet they interfere either constructively or destructively depending on their relative phase.

The wave-particle nature of light

In the previous section light was described as a form of energy and the idea that energy can be transferred between two points by waves, without movement of particles between those points, was introduced. This suggests that light is a wave motion. This is true but it is not the complete story. Light actually shows *wave-particle duality*: in some cases light exhibits the properties of a wave while in other cases it behaves more like a stream of particles (Jenkins and White, 1981; Wilson and Hawkes, 1983).

During the seventeenth century two theories about the nature of light were developed: Newton's corpuscular theory and the wave theory by Hooke and Huygens. Subsequent observations by other scientists supported the wave theory and by 1887 the work of Maxwell and Hertz had led to its acceptance as the theory of light. However, although the theory works very well when describing the interaction of light with light, such as happens in interference and diffraction, it fails completely when applied to the interaction of light with matter. The wave theory of light is unable to provide a satisfactory explanation in situations where energy is exchanged, such as in the absorption and emission of light. Another case where the wave theory breaks down is the photoelectric effect, which is the emission of electrons from the surface of a solid when irradiated. This was explained by Einstein in 1905 when he suggested that the energy in a beam of light, rather than being spread evenly, is in fact concentrated in certain regions which behave like particles. These particles were named photons. Einstein's idea of photons was derived from the work of Planck on the emission of light from hot bodies. Planck's experiments suggested that light energy is emitted in multiples of a certain minimum energy unit, a quantum, the size of which depends on the wavelength of the radiation. Thus, we have two separate theories to describe the nature of light: a wave theory and a particle theory. There is strong evidence to support both and they should be thought of as being complementary rather than contradictory. Light has dual nature. In many situations, particularly those involving the exchange of energy, the particle nature dominates the wave nature but in situations where light interacts with light the wave nature dominates.

As mentioned earlier in this chapter the laser is a source of light. The light emitted from a laser is a wave motion having a characteristic wavelength, and a velocity determined by the medium through which it is travelling. When absorption of light and its subsequent effects are of interest (e.g. in the removal of dirt layers from an artefact by laser radiation), it is more convenient to think of light as composed of localized 'bundles' of energy called photons which behave like particles. The amount of energy associated with each photon is defined by the laws of quantum physics and is related to the wavelength of the light.

The properties of light and the electromagnetic spectrum

As a wave light is characterized by a combination of oscillating electric and magnetic

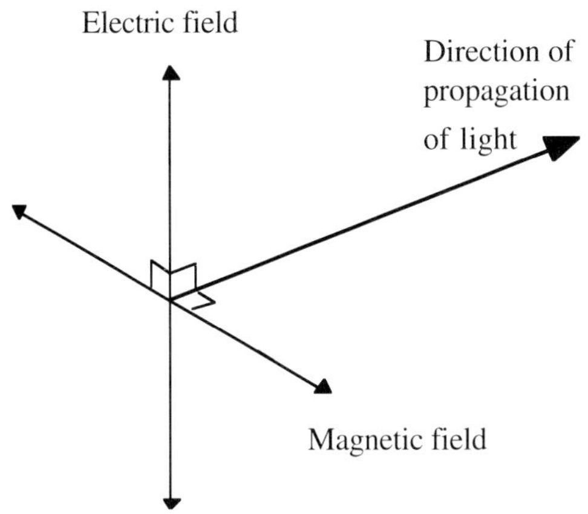

Figure 2.5. *Light as an electromagnetic wave.*

fields, i.e. light is an *electromagnetic* wave (Wilson and Hawkes, 1983). The electric and magnetic fields vibrate at right angles to each other and at right angles to the direction in which the light travels (see Figure 2.5), i.e. light waves, like water waves, are transverse waves. There are other types of electromagnetic waves similar in nature to light, differing only in wavelength: gamma rays, X-rays, microwaves and radio waves. Together they form what is known as the *electromagnetic spectrum* (see Figure 2.6). The wavelengths of visible light (which can be detected by the eye) extend between about 400 nm and 700 nm. The visible light we receive from the sun is actually composed of many different colours of light, each colour having a slightly different wavelength. This is clearly demonstrated by a rainbow: water droplets in the atmosphere act like a prism, bending each of the different wavelengths or colours slightly different amounts so that the white light is split into its red, orange, yellow, green, blue, indigo and violet components. Violet has the shortest wavelength and red the longest. Just as the ear is unable to detect sound above a certain frequency, e.g. a dog whistle, the eye is insensitive to wavelengths shorter than the extreme violet and longer than the deep red. The term ultraviolet light is used to describe radiation with a wavelength shorter than that of visible light down to a wavelength of about 30 nm. The ultraviolet portion of the sun's rays is responsible for sunburn. Between about 30 nm and 0.1 nm is the region of X-rays and at shorter wavelengths still are the gamma rays produced by radioactive substances. On the other side of the spectrum, at the long wavelength side of the visible, is the infrared (produced by hot bodies) which merges into microwaves at a wavelength of about 3×10^5 nm (0.3 μm). At even longer wavelengths, in the region between about 3×10^8 nm (0.3 m) and 1×10^{14} nm (100 km) are radio

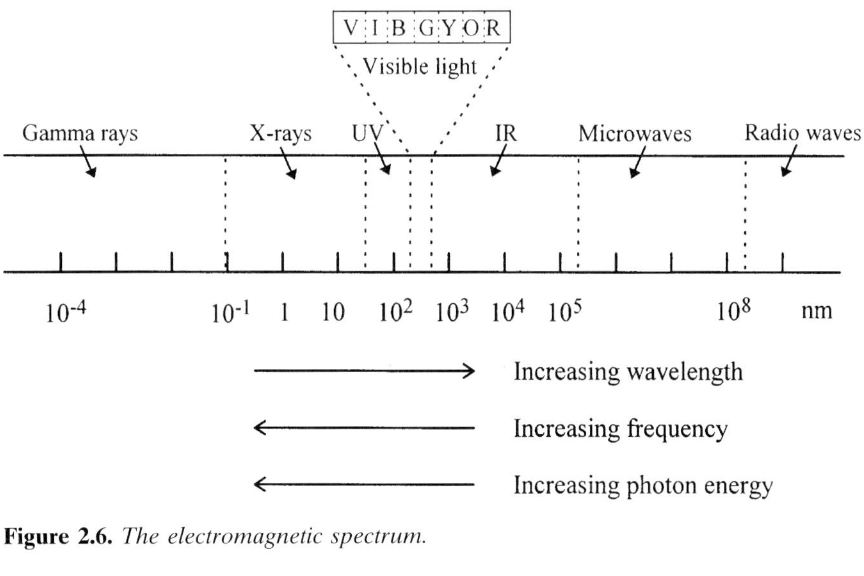

Figure 2.6. *The electromagnetic spectrum.*

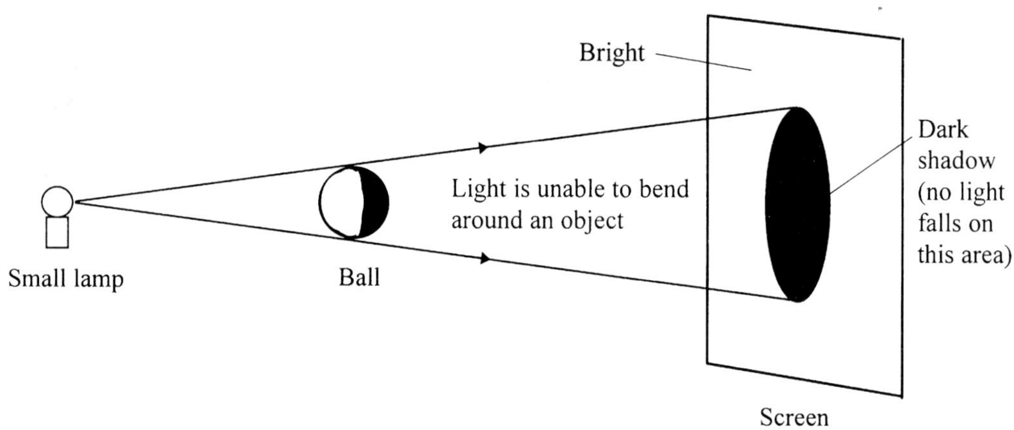

Figure 2.7. *Shadow due to a small source of light.*

waves which are used for broadcasting. As wavelength decreases the energy of the photons associated with the electromagnetic waves increases; photons in the ultraviolet are more energetic than photons in the infrared.

The dividing lines between the different regions of the spectrum are roughly fixed by the methods employed to generate and detect the different types of radiation. They are therefore not precise and very often overlap each other. Visible light covers only a very small part of the electromagnetic spectrum. Conventionally the term light is extended to include the ultraviolet and infrared portions of the spectrum. Lasers produce radiation within this region of the spectrum.

Light travels extremely quickly. In ancient times astronomers believed the speed of light to be infinite, with light from distant stars being instantaneously observable at all points in the universe. In 1600 Galileo, after failing to measure the speed of light, concluded that it must be infinite. However, in 1849 the French physicist Fizeau became the first person to measure its speed successfully (Jenkins and White, 1981), and since then many more experiments have been carried out improving the accuracy of the result. The speed of light in air is now known to be approximately 300 000 km/s – roughly equivalent to eight times around the earth in a single second! The speed of light depends on the medium through which the light is travelling, for example, the speed of light in water is approximately three-quarters and in glass two-thirds of the speed of light in air. All electromagnetic waves, from gamma rays to radio waves, are believed to travel at the same speed in a vacuum.

One other important property of light is that it travels in straight lines. This is why objects can be made to cast sharp shadows. If, for example, a small ball is placed between a light source and a screen then a shadow of the ball will be cast onto the screen as the light is unable to bend around the ball (see Figure 2.7). The same principle is responsible for an eclipse of the sun by the moon.

How a laser works

So far we have seen that light is a form of energy that exhibits both a wave-like and a particle-like nature. As a wave, light propagates with a transverse motion at a speed of about 300 000 km/s. The term *light* includes the ultraviolet, visible and infrared parts of the electromagnetic spectrum. In terms of particles, light consists of localized packets of energy, photons, the energy of which depends on the wavelength of the light. There are many sources of light: the sun, the tungsten filament light bulb, the fluorescent lamp and the glow-worm are all examples. These sources generate light in different ways but the light emitted does have several properties in common: it spreads out quickly from the source; it is composed of several wavelengths, and it is emitted in an entirely random manner (there is no fixed phase relationship between the emitted waves). The laser, however, is a unique source, producing a highly directional, 'pure' form. Within a laser, light is generated in an ordered way so that the light waves can be amplified to produce an intense beam. In order fully to appreciate the special properties of laser radiation it is helpful to understand how a laser works. To do this it is necessary to look at the processes involved on an atomic scale.

Emission and absorption of radiation

All substances are made up of atoms, ions or molecules, the basic building blocks of matter. An atom consists of a core (nucleus), which contains a fixed number of protons (positively charged particles), neutrons (particles without charge), and electrons (negatively charged particles). In an atom the number of electrons matches the number of protons so that there is no net charge. The electrons occupy fixed orbits around the nucleus (see Figure 2.8). An ion is an atom with an excess of charge, i.e. an atom which has either gained or lost one or more electrons, and a molecule is a group of atoms held together by attractive forces. From here on, the term *atomic system* will be used to describe a system made up of atoms, ions or molecules. An atom has a certain amount of energy associated with it. In its normal unexcited state this energy is known as the *ground state* energy. If an atom gains energy from external sources, such as light or heat, then the total amount of energy associated with the atom will increase and the system is now described as being in an *excited state*. The laws of quantum mechanics state that the energy of an atomic system can only take certain discrete values. When electromagnetic radiation interacts with such a system transitions occur between these discrete energy levels and energy will be either absorbed or emitted at a wavelength characteristic of the material. Transitions between discrete energy levels which describe the electron orbits in an atom often participate in the lasing process. However, transitions between other energy states, such as those associated with the

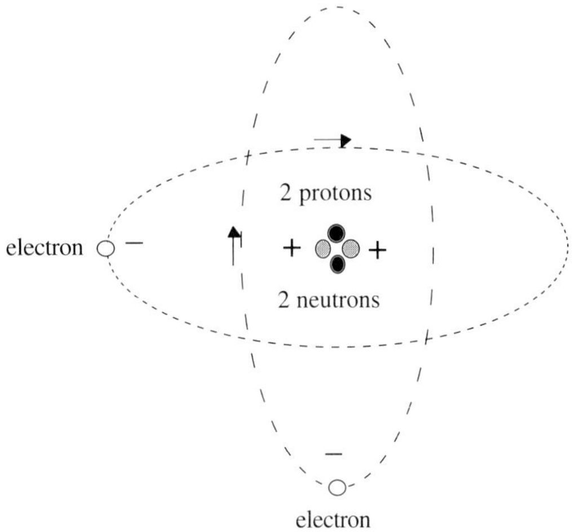

Figure 2.8. *Schematic representation of a helium atom.*

rotation of molecules, can also be used to achieve laser action. Lasing action has been achieved in gases, liquids and solid materials. The range of wavelengths available from lasers is therefore very wide, extending from the ultraviolet to the infrared. A simplified model of the atomic system consists of many identical, two-energy-level atoms, each with a ground state, E_1, and an excited state, E_2, (Kestenbaum, 1972; Wilson and Hawkes, 1983b). One such atom is shown schematically in Figure 2.9.

Consider a simplified two-level energy system. An atom in its ground state in the presence of photons of energy (E_2–E_1) is raised to its excited state by absorbing a photon (see Figure 2.10a). In time an atom already in an excited state will return to its more stable ground state by emission of the

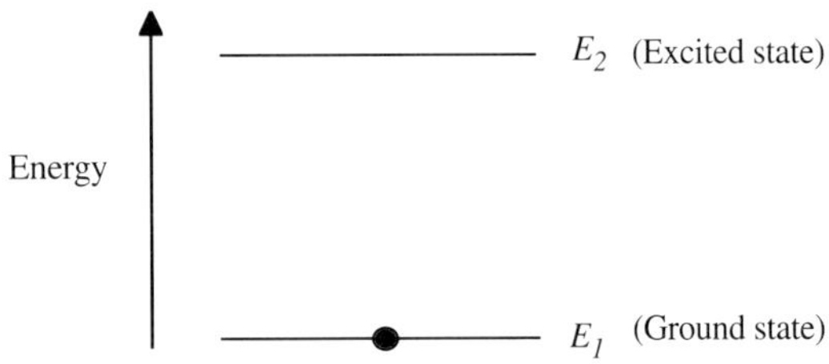

Figure 2.9. *A two-level energy system showing an atom in its ground state.*

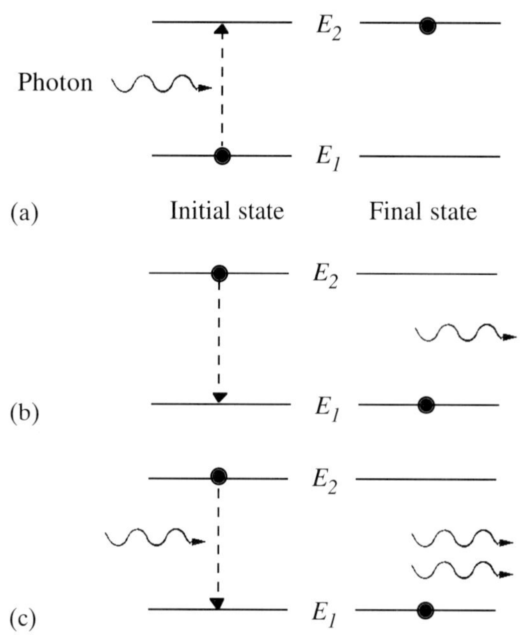

Figure 2.10. *The process of absorption and emission of light for a two-level atomic system.*

excess energy. The energy can be emitted in several forms, e.g. heat or light. If the energy is emitted as a photon then the wavelength of the light is determined by the value of $(E_2–E_1)$. The emission process can occur in two ways. The most likely way is for the atom simply to drop back down to its lower level in an entirely random manner. This process is known as *spontaneous emission* (see Figure 2.10b). Consider millions of atoms undergoing this process simultaneously. The random nature of the process means that light is emitted in many directions without any phase relationship between the waves. Light generated in this manner, such as that by the sun, is said to be *incoherent.* The emission process can also occur by *stimulated emission* in which an excited atom is 'triggered' to undergo the transition to the lower energy level by the presence of photons of energy $(E_2–E_1)$ (see Figure 2.10c). Light emitted by this process is identical to the 'stimulating' wave, that is, it has exactly the same wavelength, phase and direction of propagation. It is said to be *coherent.* This means that the amplitude of an incident wave can grow as it passes through a collection of atoms, i.e. the process of stimulated emission can lead to amplification of an incident wave. Usually stimulated emission is not observed since the probability of spontaneous emission occurring is much higher. In order to amplify a light wave by stimulated emission it is, therefore, necessary to increase the rate of this process relative to the rates of spontaneous emission and absorption.

Population inversion

The rate of stimulated emission is proportional to the energy density of the stimulating field at the appropriate wavelength, i.e. the number of photons with energy $(E_2–E_1)$, and the number of atoms, or population, of the excited state, N_2. In order to increase the rate of stimulated emission in relation to the rates of spontaneous emission and absorption, both the energy density and population of the excited state, N_2, relative to the population of the lower level, N_1, must be increased. It can be shown that a necessary condition for the amplification of a beam of light as it passes through a collection of atoms is $N_2>N_1$ (Wilson and Hawkes, 1983). The situation where more atoms are in the higher energy level is known as a *population inversion*. It is a non-equilibrium distribution of atoms among the energy levels of an atomic system. For a system in thermal equilibrium the lower energy levels are more densely populated than the higher energy levels (see Figure 2.11a). In fact, for an energy difference $(E_2–E_1)$ large enough to give visible radiation the population of the upper level is almost negligible compared to the population of the lower level. In order to create a population inversion (Figure 2.11b) a large amount of energy must be supplied to the system to excite atoms into the upper energy level. This excitation is achieved by a process known as *pumping*. Much of the development work of lasers has concentrated on the problem of supplying pumping energy to a laser system.

Stimulated absorption is one of the methods used for pumping. The energy levels taking part in the lasing transition are pumped by intense irradiation of the system, that is, a pumping source is chosen which supplies a high concentration of photons with sufficient

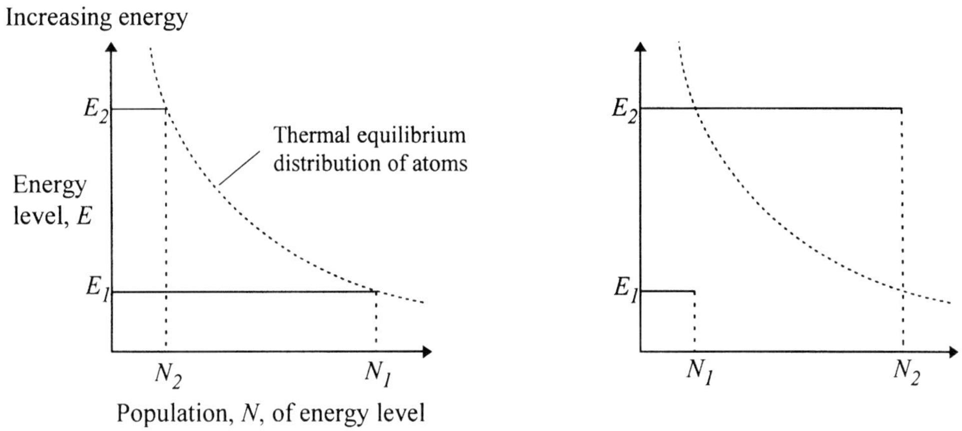

Figure 2.11. *Graph showing the population of different energy levels in a two-level atomic system. N_1 represents the population of level E_1.*

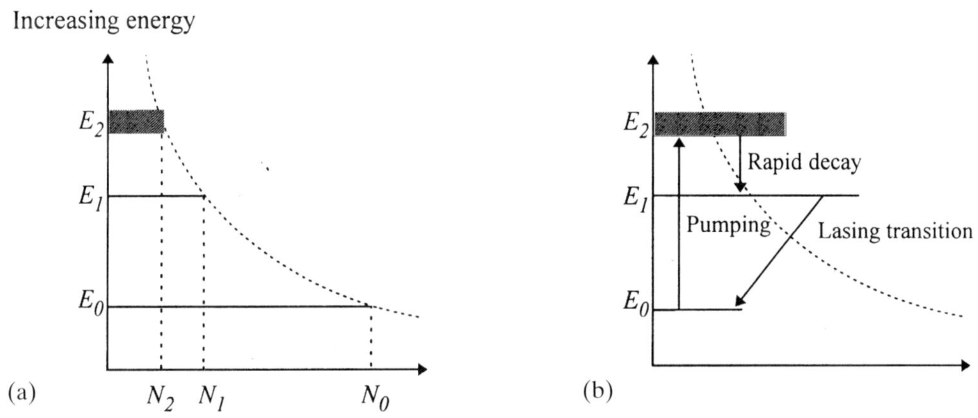

Figure 2.12. *Graph showing the population of energy levels in a three-level system.*

energy to bring about a transition from level E_1 to E_2. With a two-level energy system the probabilities of stimulated emission and absorption become equal as soon as atoms are excited to the upper level. The best that can be achieved, even with intense pumping, is equality of the populations: $N_2=N_1$. It is, therefore, necessary to use materials with either a three- or four-level energy system.

Consider the three-level system shown schematically in Figure 2.12. In normal conditions the population of the energy levels follows the equilibrium distribution: $N_0>N_1>N_2$ (Figure 2.12a). Intense illumination of the system raises atoms from the ground state E_0 to the excited state E_2. In a laser E_2 is unstable and the atoms quickly decay to the more stable level E_1 (according to the laws of quantum physics). Atoms collect in level E_1 (as a result of its relative stability to E_2) until eventually a population inversion is created between E_1 and E_0. The lasing transition, the stimulated emission of light, takes place between these two levels (see Figure 2.12b). Ideally, the transition from E_2 to E_1 should be rapid so that there are always available states at E_2 and the transition from E_1 to E_0 should be very slow to allow a large build-up of atoms in E_1. Also, the level E_2 should consist of many closely-spaced energy levels so that a wide part of the spectral range of the pumping radiation can be used. This increases the pumping efficiency. A three-level system, such as the ruby laser, requires powerful pumping as the final level of the lasing transition is the ground state. Four-level systems, such as the Nd:YAG laser, have much lower pumping requirements.

Pumping can be achieved in a variety of ways. In addition to optical methods an electrical discharge, electron bombardment, the release of chemical energy and the passage of current can all be used. The energy systems of laser media (solid, liquid or gas) are often very complex and are beyond the scope of this book but can usually be simplified to either a three- or four-level system.

Inside a laser

Although lasers come in many shapes and forms they all have three important features in common: a lasing medium, a pumping source and an optical cavity. The lasing medium provides the atomic system for lasing and hence defines the wavelength of laser radiation generated. The pumping source provides the medium with sufficient energy to attain a population inversion so that stimulated emission can take place. The optical cavity forms a feedback system which allows the amplitude of the light wave to grow sufficiently to overcome losses in the system. Inside a laser the lasing medium is positioned between two mirrors to form the optical cavity. The initial stimulus for lasing is given by spontaneous transitions between the energy levels participating in the laser transition in which the emitted photons travel along the axis of the system. The light is amplified as it passes through the medium, due to stimulated emission, and is 'fed back' by the mirrors. Any photons emitted in directions other than along the axis are quickly lost. The light generated by the laser therefore takes the form of a well-collimated beam, and amplification, during the multiple passes that occur within the cavity, can become substantial. One of the mirrors is made as reflective as possible while the other is made only partially reflecting. Transmission through this mirror provides the output from the laser.

A schematic representation of a pulsed Nd:YAG laser (currently the most common laser used in conservation) is shown in Figure 2.13. The lasing medium is a rod, several centimetres long, of yttrium aluminum garnet (YAG) with neodymium (Nd) ions present as an impurity. The Nd ions provide the energy levels for the lasing transition and the pumping, while the YAG crystal lattice modifies the transition probabilities between

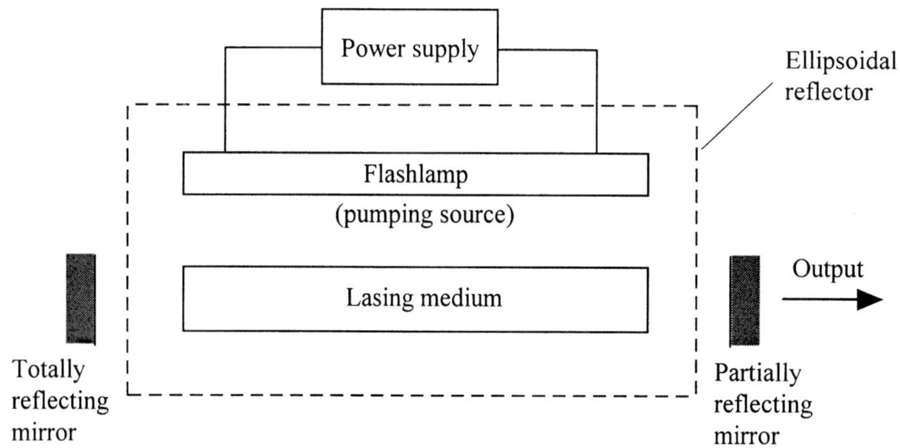

Figure 2.13. *Schematic representation of a flashlamp-pumped Nd:YAG laser.*

some of the energy levels in the Nd ion, making conditions more favourable for lasing. The pumping source is a flashlamp, often xenon, which provides a very intense flash of light in much the same way as a camera flash. The flashlamp and laser rod are housed inside an ellipsoidal reflector so that as much of the pumping radiation as possible falls onto the rod. This increases the efficiency of the laser. Since the pumping flash lasts for only a short time (typically 1 ms) the output from the laser is in the form of a pulse (typically 0.5 ms). By replacing the flashlamp with quartz-halogen lamps it is possible to obtain a continuous output from the laser. During operation the pumping source generates a large amount of heat which can cause the laser rod to become very hot. Cooling, in the form of air or water passed across the surface of the rod, is therefore necessary to avoid damage to the rod and to allow operation at increased repetition rate.

In a Nd:YAG laser the lasing transition results in emission of coherent radiation at a wavelength of 1.06 μm in the near infrared part of the electromagnetic spectrum. In conservation the pulse length is usually required to be much shorter than the duration of the flashlamp: nanoseconds rather than milliseconds. This is achieved by a process known as *Q-switching*, in which a shutter is used to isolate the lasing medium from the cavity. By doing this the pumping energy can be stored in the excited states of the lasing medium until a large population inversion has been achieved. The shutter is then opened and closed very quickly so that all of the energy is released in a very short pulse. In the ordinary pulsed mode the peak power (pulse energy divided by pulse duration) is typically of the order of kilowatts. When the pulse is Q-switched approximately the same amount of energy is emitted in a much shorter time (10–100 ns); the peak power of this 'giant' pulse is typically of the order of megawatts (see Figure 2.14).

In summary, the basic conditions which must be satisfied for laser operation are:

1. an active medium which emits radiation in the appropriate region of the electromagnetic spectrum;
2. the energy levels associated with the lasing transition must be suitable for pumping so that a population inversion can be achieved, and
3. optical feedback at the ends of the medium to form a resonant cavity.

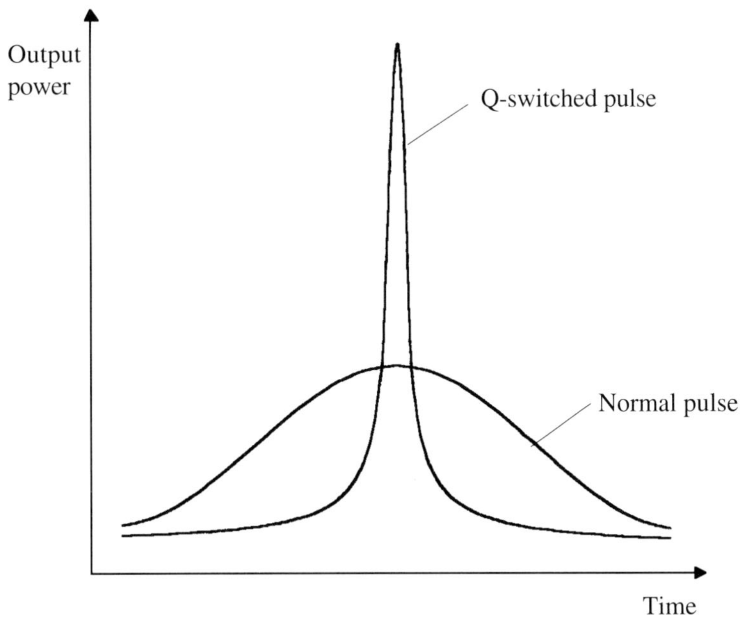

Figure 2.14. *Schematic representation showing the difference between a Q-switched laser pulse and a normal laser pulse (not to scale).*

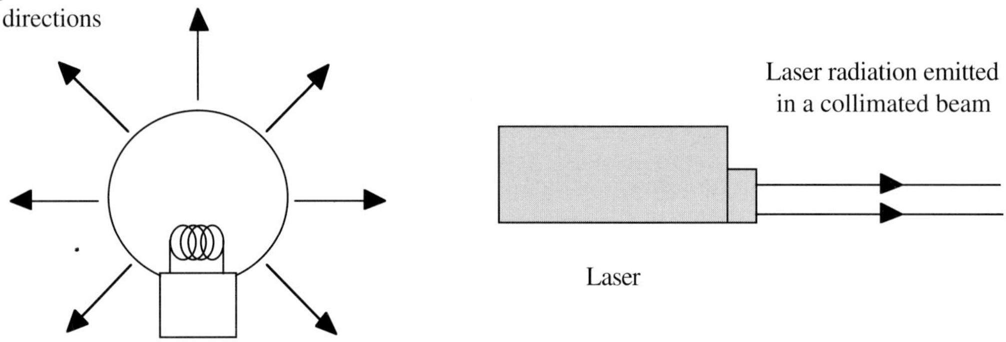

Figure 2.15. *The highly directional nature of laser radiation.*

Properties of laser radiation

Laser radiation is a unique form of light. It possesses special properties, such as directionality, brightness, purity and in some cases tunability, which make it suitable for a wide range of applications (Wilson and Hawkes, 1983). It should be noted that not all lasers exhibit these properties to the same extent. This will often restrict the choice of laser for a particular application.

Directionality

Unlike a normal light source, such as a filament lamp, which emits light in all directions, laser radiation is emitted in a highly directional collimated beam with a small angle of divergence (see Figure 2.15). The fact that laser radiation spreads out only very slightly, often as little as a few millimetres over a distance of 10 m, means that it is straightforward to collect the energy in a laser beam and focus it into a small area even at relatively large distances from the laser. Efficient collection of the energy from a normal light source is almost impossible. The low divergence of laser beams has led to applications in alignment (e.g. the helium neon (HeNe) laser) while the ability to focus large amounts of energy very precisely into a small area has led to applications in drilling, cutting, welding and other surface treatments (e.g. the carbon dioxide laser). Typical divergences for lasers are given in Table 2.1.

Table 2.1 Typical values of beam divergence for several types of laser (beam divergence of 1 mrad corresponds to an increase in beam diameter of 10 mm over a distance of 10 m)

Type of laser	*Wavelength/nm*	*Operating mode*	*Divergence/mrad*
HeNe	633	Continuous	0.8
Nd:YAG	1064	Q-switched	3
Carbon dioxide	10.6×10^3	Pulsed	2
Ruby	694	Q-switched	10

Brightness

The *brightness* of a source can be defined as the power emitted per unit area per unit solid angle where the solid angle is that defined by the cone into which the beam spreads. Since the laser emits large amounts of power into a well-collimated beam it is a source of great brightness. The brightness of a laser exceeds that of any other light source: it is brighter than the sun by several orders of magnitude! The high brightness and low divergence of a laser beam mean that large amounts of power per unit area can be delivered to a target which is essential for many applications including laser cleaning. It also means that a stray laser beam can cause serious damage to the eye and appropriate eyewear must, therefore, be worn at all times when using a laser.

Purity

The laser is a source of extremely monochromatic, highly coherent light. Laser radiation is a very 'pure' form of energy. Unlike many other sources, such as the sun, which emit radiation over a continuous spectrum from the ultraviolet to the infrared, a laser emits radiation with a very narrow linewidth, i.e. the output from a laser can be considered as monochromatic (see Figure 2.16). This means that by choosing an appropriate wavelength the laser can become an extremely selective tool. The amount of energy that a particular target absorbs from a laser beam depends to a large extent on the wavelength of the beam. In order carefully to remove a surface layer from a substrate it is desirable that the surface layer absorbs the incident radiation very strongly but the substrate absorbs only very weakly, if at all. By selecting a light source that emits radiation at one wavelength only it is often possible to satisfy these criteria. If the light source emits radiation over a continuous spectrum the chances are that the substrate will also absorb strongly over a certain region of the spectrum, making selective removal impossible.

Laser radiation has a high degree of coherence, in both time and space, i.e., light at different points, both transverse to and along the direction of travel of the beam, has a fixed phase relationship. A high degree of coherence is especially important in applications where the laser beam is split into two or more beams which travel along different paths before recombining, such as interferometry (measurement of small distances and movements) and holography. Continuous gas lasers, such as the HeNe laser, have very good coherence.

If the output from a laser is examined at very high resolution then it becomes clear that it actually consists of a small number of very narrow spectral lines which are a result of standing wave patterns set up within the optical

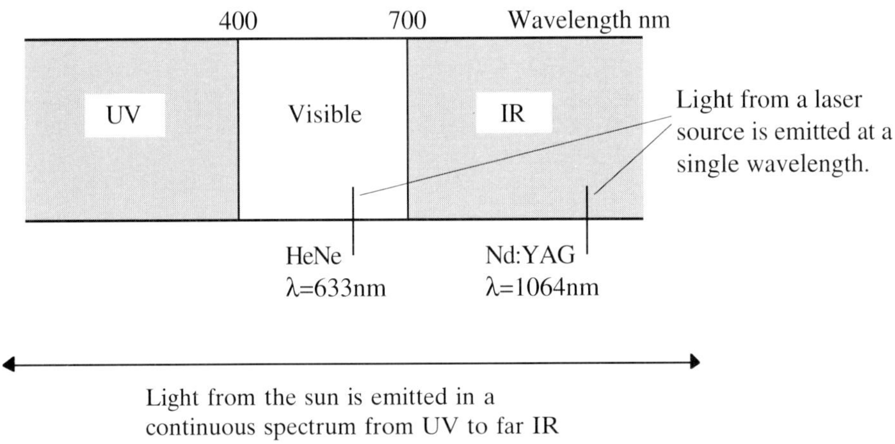

Figure 2.16. *Comparison between the spectra of light emitted from the sun and a laser.*

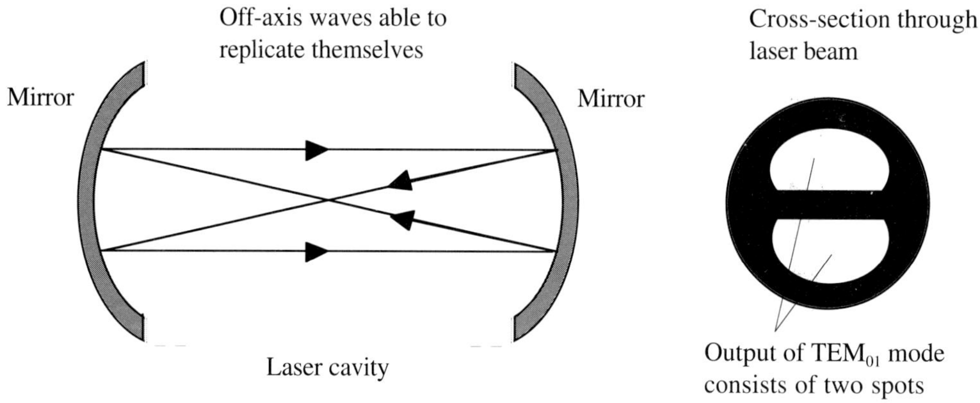

Figure 2.17. *Formation of a transverse electromagnetic (TEM) mode.*

cavity (Wilson and Hawkes, 1983). The standing waves are formed between the mirrors in the same way as standing waves develop on a string or in an organ pipe. These spectral lines, or axial *modes*, all contribute to a single spot in the output. In addition to axial modes any real laser cavity also has *transverse modes* which are formed by waves travelling just off-axis that are able to replicate themselves (see Figure 2.17). Transverse modes give rise to variations in the energy distribution in the laser beam. The number of transverse modes oscillating depends on the size of the aperture of the cavity. The greatest spectral purity and coherence are achieved by a laser operating in a single axial and single transverse mode. This can be achieved by reducing the length of the cavity and by placing an aperture within it. The high degree of purity achievable with certain types of laser has led to applications in basic scientific research including photochemistry, atomic, molecular and Raman spectroscopy, and also in communications. Operation of a laser in multimode form leads to a fall-off in spectral purity and coherence but a considerable increase in power. In cleaning applications the Nd:YAG laser is usually operated in multimode form since the need for power usually outweighs the need for absolute spectral purity.

Tunability

Some lasers can be tuned to emit radiation over a range of wavelengths. Dye lasers can often be tuned over a large region of the visible spectrum which has opened up applications in photochemistry and high resolution spectroscopy. Harmonic generation and parametric amplification (Wilson and Hawkes, 1983) have opened up new possibilities in solid state lasers. By placing an anisotropic crystal, such as potassium dihydrogen phosphate (KDP), into the path of a Nd:YAG laser beam it is possible to generate radiation at its harmonics: 532, 355 and 266 nm. In this case the same laser system can provide outputs in the infrared, visible and ultraviolet. Multiple laser systems are also available in which the second harmonic output from a Nd:YAG laser (532 nm) is used to pump a dye laser, thereby further increasing the number of wavelengths provided by a single system. Disadvantages of these techniques include a loss in power during the conversion process which is often accompanied by a degradation in beam quality. A laser system that can provide spectrally pure, coherent radiation in tunable form, from the ultraviolet to the infrared without any loss in output power, unfortunately does not yet exist.

Summary

A useful analogy illustrating the special properties of laser radiation is to think of light waves as people! Light from a conventional incoherent source can be compared to the hustle and bustle of a crowded shopping centre with people rushing around in all directions, oblivious to what is going on around them. Light from a laser is much more orderly. It can be compared to a column of marching soldiers, all wearing the same uniform, all moving in the same direction, marching precisely in step with each other.

Different lasers exhibit these properties to different extents. In selecting a laser for a particular application it is necessary to consider the relative importance of properties such as divergence, spectral purity and coherence of the beam in addition to the more obvious characteristics: wavelength, pulsed or continuous and power. For cleaning applications, once the wavelength (for our purposes the laser beam can be considered to be monochromatic) and pulse length have been selected, the most important consideration is the power of the laser; each pulse must be sufficiently energetic to remove dirt from a substrate. Since a high degree of coherence is not important the laser can be operated in multimode form. Beam divergence is also an important consideration since it is desirable that as much of the energy in the beam as possible should reach the target.

Types of laser and their applications

Laser action has been achieved in hundreds of different substances: solid crystals such as ruby, gases such as carbon dioxide and liquids such as organic dyes. The wavelengths generated by such a wide range of substances extend from the ultraviolet to the far infrared (see Figure 2.18). The diverse material properties of the lasing media mean that the radiation emitted has distinct characteristics: degree of spectral purity, degree of coherence, level of divergence and power. Radiation emitted from a single mode HeNe gas laser has a high degree of spectral purity and coherence lending itself to applications in holography, whereas the radiation emitted from a high power multimode Nd:YAG laser, although not possessing the same 'purity', does have sufficient power to cut through a sheet of steel. Such a wide range of characteristics has led to an equally diverse range of applications. As described in Chapter 1, lasers are now commonplace in hospitals, museums, industry, the armed forces and basic scientific research. Lasers have even been used to cut through diamonds, initiate thermonuclear reactions and measure the distance to the moon! Currently lasers are used to cut, drill, weld, engrave, coat, mark, read, measure, align, display and clean. This list is by no means exhaustive but does serve to illustrate the extremely wide range of tasks that lasers are now employed to carry out. Obviously a single laser system cannot carry out all of these tasks.

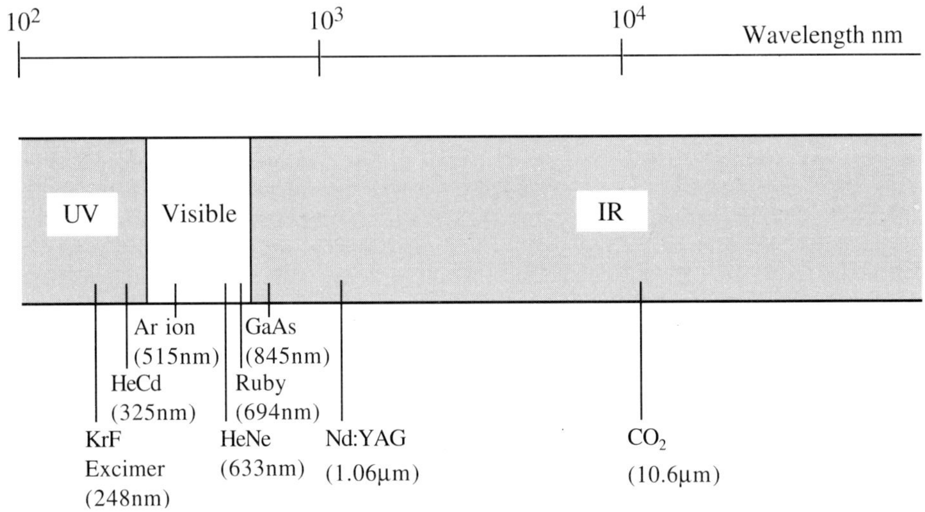

Figure 2.18. *Some of the more common types of laser.*

Selection of the most appropriate laser for a given task is critical. Some of the more commonly available lasers currently being used in conservation work and research are described here. They are divided into three groups: solid state, gas and dye lasers.

Solid state lasers

The active medium of a solid state laser usually consists of a crystalline array of atoms with impurity atoms introduced during growth of the crystal. Solid state lasers are generally rugged, straightforward to maintain and capable of delivering high peak powers. Ruby and Nd:YAG lasers are two examples of such lasers. A single crystal of synthetic ruby was the first material to demonstrate lasing action (Maiman, 1960). The ruby is a crystal of corundum (Al_2O_3) doped with chromium ions in the concentration 0.05% by weight. The chromium ions give ruby its familiar pink colour and provide the energy levels for the lasing transition, giving visible red light at a wavelength of 694 nm. Being a three-level system the ruby laser requires intense pumping to achieve a population inversion. This is usually achieved by discharging a capacitor bank (storing a few thousand joules of energy) into a flashlamp. The efficiency (ratio of laser energy out to total electrical energy in) of a ruby laser is relatively low, typically less than 1%. Q-switched ruby lasers can produce peak powers in the range of several hundred megawatts lasting 10–20 ns, while further modification can give rise to peak powers of several gigawatts in pulses lasting just a few picoseconds.

The active medium of a Nd:YAG laser is a crystal of yttrium aluminum garnet doped with neodymium ions. Principles of operation are similar to the ruby laser, although the Nd:YAG laser, being a four-level system, has much lower pumping requirements. Its efficiency is typically 2%. The Nd:YAG laser is now much more widely used than the ruby laser. It has been described in more detail earlier in this chapter. Output is in the near infrared at a fundamental wavelength of 1.06 μm and can be in continuous or pulsed form depending on the pumping source. Q-switching provides pulses in the nanosecond range and other techniques can reduce the pulse duration to a few picoseconds. Nd:YAG lasers can provide several hundred watts in continuous form but most are pulsed for high peak power. Pulse energies of several joules can be obtained and Q-switching can produce peak powers of 100 MW at repetition rates of up to 100 Hz. Such lasers are used in industry for marking and engraving and, since the early 1990s, have found applications in conservation where the wavelength, high peak powers, rugged design and reliability are suited to the controlled cleaning of many artworks. As mentioned earlier in this chapter, by passing the fundamental wavelength of a Nd:YAG laser through a non-linear crystal, outputs in the green and ultraviolet parts of the spectrum can be obtained. This process, known as harmonic generation, is accompanied by a fall-off in output energy and degradation in beam quality. However, further development of such a system could possibly lead to an extremely versatile tool, offering outputs in the near infrared, visible and ultraviolet parts of the spectrum.

Gas lasers

Gas lasers are the most commonly used type of laser. They range from the low power helium-neon (HeNe) laser used in teaching laboratories to the high power carbon dioxide laser which today plays an important role in many industrial manufacturing techniques. The HeNe laser was the first gas and the first continuous laser. Its active medium is a mixture of helium and neon gases, typically in the ratio 10:1. Neon provides the energy levels for the lasing transition and helium plays an important role in the excitation process, helping to transfer energy from a direct current discharge to the neon. There are a number of lasing transitions available within the gas and the most commonly available wavelengths from a HeNe laser are 633 nm (red) and 543 nm (green). Although the output power of a commercial HeNe laser is typically only a few milliwatts (up to a maximum of about 100 mW) it has found applications in a wide range of areas since its radiation has very low divergence, a high degree of coherence and is extremely monochromatic. Typical applications are in holography, scanning systems and alignment systems. In conservation the HeNe laser often provides an aiming beam for the invisible Nd:YAG beam during cleaning.

The carbon dioxide (CO_2) laser is the most important gas laser from both an industrial and scientific viewpoint. Lasing transitions occur between the discrete energy levels provided by vibration and rotation of the carbon dioxide molecules. Radiation is emitted at 10.6 μm in the far infrared in either pulsed or continuous form with high efficiency (up to about 30%). The range of power levels available from the many different designs is extremely wide: continuous power outputs vary from a few watts up to as much as 100 kW, and has given rise to many important industrial applications. Carbon dioxide lasers are used to drill and cut metals, polymers and ceramics. Power levels required for cutting range from tens of watts for thin plastics and ceramics to over 1000 W for metals. Other forms of carbon dioxide laser are capable of providing energies of up to 150 J, in pulses of several nanoseconds up to 100 ms at repetition rates of 0.1–1000 Hz. These lasers are used in applications, such as marking, where a large beam of high pulse energy is required. The processing of metals with a carbon dioxide laser requires high power levels since most metals reflect very strongly at 10.6 μm. The feasibility of using low energy pulses (of about 1 J) to clean metal sculpture is currently being investigated (Cottam, 1997).

Excimer lasers provide a source of coherent ultraviolet radiation. An excimer is a molecule formed by the association of an excited atom with another atom in its ground state. The lasing transition is between the excited and ground states of the excimer. The bond between the atoms is very strong but the lifetime of the excited state is very short. As soon as the excimer reaches the ground state it dissociates, effectively reducing the population of the lower lasing level. This means that it is relatively easy to achieve a population inversion. Pumping is usually achieved with an intense electron beam source. Depending on the gas used average powers of over 100 W are available. A commonly used excimer laser is the krypton fluoride (KrF) laser which emits radiation at 248 nm with an efficiency of 10–15%. The photons in an excimer laser beam are sufficiently energetic to interact directly with the chemical bonds in polymers. This has led to applications in polymer processing, particularly in the microelectronics industry, since less thermal damage occurs allowing sharper edges to be cut. In conservation, this property of excimer laser radiation is being exploited in the cleaning of paintings where the generation of heat in a surface can have devastating effects on the pigment (see Chapter 6).

Dye lasers

Liquid media have several advantages over solid and gas laser media. They are easier to prepare than solids, which require a high degree of optical homogeneity, and have a higher density of active atoms than a gas. The active medium of a dye laser is an organic dye dissolved in a solvent. When the dye is excited by short wavelength radiation it emits light at a longer wavelength, i.e. it fluoresces. The emission spectrum of a dye is relatively broad. By replacing the output mirror of the cavity with a diffraction grating the dye laser becomes a tunable source. The dye rhodamine 6G, for example, has a tuning range from 570 to 660 nm with a peak output at about 590 nm. Using about six dyes it is possible to obtain wavelengths covering the whole of the visible spectrum. Pumping is achieved by optical means: flashlamps and other lasers. Dye lasers are usually operated in pulsed mode, the pulse length depending on the type of pumping source used. Pulse energies available from commercial dye systems are typically a few hundred millijoules from a flashlamp-pumped system, giving peak powers of several hundred kilowatts. Continuous operation is possible if a continuous pumping source is used. Output powers are typically tens of milliwatts and are dependent on wavelength. The dye laser is an extremely useful tunable source of coherent radiation in applications requiring relatively small pulse energies and favouring wavelength selectivity, e.g. spectroscopy. Commercial systems designed for medical use are now available in which a mini dye laser can be fitted to a frequency-doubled Nd:YAG source, thereby extending the range of wavelengths available to a Nd:YAG system. It will be interesting to see whether such 'multi-laser' systems will enhance the versatility of cleaning systems used in conservation.

References

Cottam, C. (1997) *Restauratorenblatter*, Proceedings of LACONA I (Crete, October 1995), International Institute for Conservation of Historic and Artistic Works, Verlag Mayer, Vienna

Jenkins, F.A. and White, H.E. (1981) *Fundamentals of Optics*, McGraw-Hill

Kestenbaum, A. (1972) Laser fundamentals. In *Lasers in Industry* (S.S. Charschan, ed.) pp. 4–14, Van Nostrand Reinhold Company

Maiman, T.H. (1960) Stimulated optical radiation in ruby masers. *Nature*, **187**, 493

Wilson, J. and Hawkes, J.F.B. (1983) *Optoelectronics: An Introduction, Prentice Hall International*

3

Removal of surface layers by laser radiation

Martin Cooper

Ever since its first demonstration in 1960, material scientists have used the 'pure' concentrated energy emitted by a laser to remove surface material from solids. In 1962 a focused ruby laser was used to vaporize and excite atoms from solid surfaces (Breech and Cross, 1962). The light emitted during the process was used to characterize the elements in the surface. This technique was called Laser Microprobe Emission Spectroscopy (LMES) and was the first real application of laser ablation. The term *laser ablation* is used to describe a process in which material is ejected from a solid surface as a result of irradiation by a laser beam. Early uses of LMES included the elemental analysis of geological, biological and metallic materials. The physical processes involved in laser ablation are extremely complex. Conversion of electromagnetic energy (the laser beam) into electronic, thermal, chemical and mechanical energy at the surface gives rise to ejection of material in the form of atoms, molecules, clusters, positive and negative ions, electrons and photons (Miller, 1994a). Although the mechanisms of the process are not yet fully understood and are still being studied, laser ablation has found many important applications: industrial processes such as hole drilling and laser cutting; materials processing to produce thin films and microstructures (submicron structures have led to smaller devices for faster computers); biomedical uses including laser surgery (ablation of tissue, tumours and dental material have all been demonstrated), in particular eye surgery; laser cleaning which encompasses a wide range of areas including removal of particulates from semiconductors during fabrication, stripping of paint from aircraft and removal of soiling from works of art.

This chapter aims to give the reader an understanding of the mechanisms believed to be involved in the laser cleaning of artworks. The early part is devoted to a description of the effects caused by absorption of the energy in a laser beam. A parallel is then drawn between the introduction of lasers for cleaning in industry (particularly the microelectronics industry) and in conservation: in both situations the introduction of lasers was seen as a means of refining the cleaning process. Some of the ideas developed in industry are used to help explain laser cleaning in conservation. The latter part of this chapter is devoted to a description of the mechanisms by which the cleaning of artworks, at the 1.06 μm wavelength of the Nd:YAG laser (which is currently the most commonly used laser in conservation), is thought to proceed. The mechanisms by which ultraviolet laser cleaning proceeds are briefly described at the end of this chapter and in Chapter 6 (Lasers in the conservation of painted artworks).

Absorption, reflection and transmission

In order to produce any lasting effect on a surface laser radiation must first be absorbed. The way in which a laser beam interacts with a material depends on the beam parameters, such as wavelength and power, and the physical and chemical properties of the material. When a laser beam meets a boundary between two media, for example between air and stone, part of the beam is reflected and the remainder absorbed (assuming no transmission through the material) (see Figure 3.1). The fraction of energy reflected is known as the *reflectivity*, r, and the fraction absorbed as the *absorptivity*, a, of the surface. The condition, $a + r = 1$, is always satisfied.

The extent of absorption depends upon two main variables: the wavelength of laser radiation and the properties of the material. Consider a white piece of paper. When daylight falls on the surface each of the different wavelengths, or colours, which make up the light are reflected strongly and the paper appears white. Black type on paper appears black because each of the wavelengths is absorbed very strongly and little or none is reflected. An object appears coloured if the light absorbed at its surface is not absorbed uniformly across the visible spectrum (Hecht, 1987). For example, a surface which appears red when illuminated by daylight usually absorbs strongly in the green and blue parts of the visible spectrum, while a surface which appears green usually absorbs strongly in the red and blue parts. A red surface would, therefore, be expected to absorb a green HeNe laser beam stronger than a red HeNe beam. In the infrared and ultraviolet the situation is not quite so obvious since the eye is unable to detect these wavelengths. In the same way as in the visible part of the spectrum, however, the absorptivity of a surface at a particular wavelength depends on its chemical and physical properties. One of the reasons why most laser cleaning in conservation is carried out at the 1.06 μm wavelength of the Nd:YAG laser is that many contaminants absorb strongly at this wavelength but many substrates absorb only weakly. At ultraviolet wavelengths most surfaces absorb strongly and, consequently, there is very little discrimination between clean and dirty surfaces.

Absorptivity also depends on surface roughness. A rough, diffusely reflecting surface absorbs a greater proportion of the energy in a laser beam than a smooth specularly (like a mirror) reflecting surface of the same material (see Figure 3.2). A rough surface presents a greater surface area to the laser beam and reflected elements of the beam may undergo several interactions with the surface rather than just one (a rough surface can be thought of as

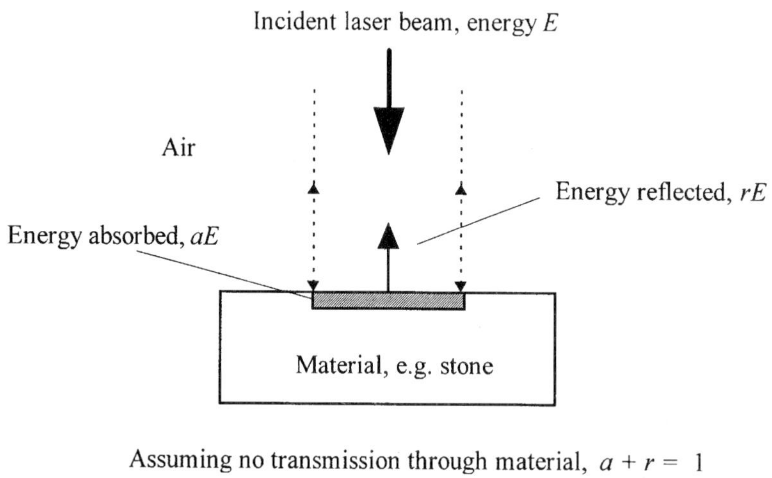

Figure 3.1. *Reflection and absorption of a laser beam at a surface.*

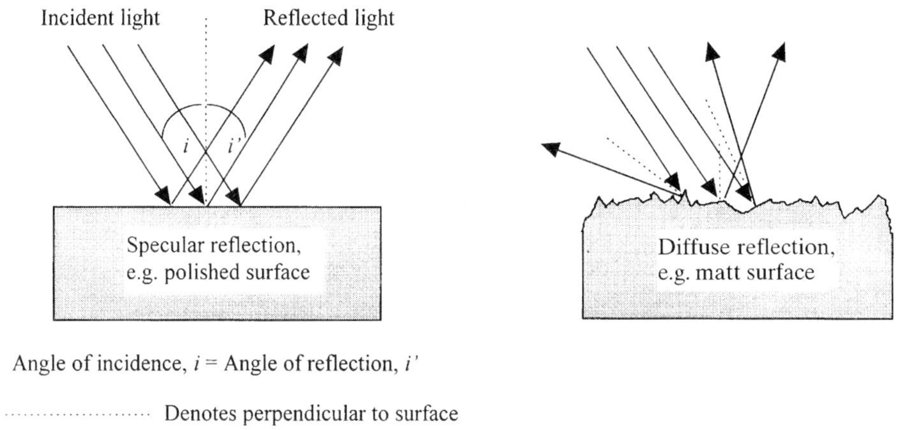

Figure 3.2. *Specular and diffuse reflection from a surface.*

being made up of a large number of small flat surfaces each one of which reflects specularly).

On an atomic scale laser radiation is absorbed in a solid by interactions with either free electrons (e.g. in a metal), bound electrons (e.g. in an insulator) or lattice vibrations. By absorbing a photon a free electron gains kinetic energy, a bound electron may be excited to a higher energy state as might a crystal lattice (energy associated with lattice vibrations can only take certain discrete values in the same way as energies of electrons bound to an atom). At high intensities interactions with electrons or the lattice may involve more than one photon. In a metal absorption occurs within a very thin surface layer, typically several nanometres. In a non-metal the radiation penetrates further into the sample, typically to a depth of several microns.

Effects due to absorption of laser radiation

The energy of a photon is inversely proportional to the wavelength of laser radiation, i.e. as the wavelength decreases the photon energy increases (see Figure 2.6). This means that ultraviolet photons are more energetic than infrared photons, e.g. a photon emitted by a KrF excimer laser at a wavelength of 248 nm is approximately four times as energetic as a photon emitted by a Nd:YAG laser at a wavelength of 1064 nm. The energy of an ultraviolet photon is higher than the energy required to break most covalent chemical bonds (Table 3.1). This difference in photon energies leads, upon absorption, to different effects in the surface of a material.

At short wavelengths, e.g. in the ultraviolet, photon absorption in covalently-bonded materials can lead to an electronic transition to a state above the dissociation limit of a molecule, causing it to split into its constituent atoms as the inter-atomic bond is broken. This chemical bond-breaking is achieved without generation of heat. At longer wavelengths, e.g. in the visible and infrared, photon energies are too low to break bonds directly and the energy is converted into heat. Once laser radiation has been absorbed in a solid, conversion of the energy

Table 3.1 Average covalent bond energies compared to photon energies at near infrared and ultraviolet wavelengths (1 eV = 1.6 × 10^{-19} J)
Photon energy (λ = 1.06 μm) = 1.2 eV
Photon energy (λ = 248 nm) = 5.0 eV

Type of covalent bond	*Bond energy/eV*
C–C	3.6
C–O	3.7
C–H	4.3
O–H	4.8
C=C	6.4

into heat is governed by collision processes in which excited electrons give up energy by means of collisions with the lattice, other electrons, ionized impurities and defects (Gagliano and Zaleckas, 1972). Since the mean time between collisions (10^{-14}–10^{-12} s) is much shorter than the duration of a Q-switched pulse (10^{-9}–10^{-8} s) or a normal-mode pulse (10^{-4}–10^{-2} s) we can assume that the laser energy is instantaneously converted to heat at the point where it is absorbed. A laser emitting low energy photons (long wavelength radiation) can, therefore, be considered as a heat source, with absorption of the beam inducing a temperature rise on and just below the surface. The advantage of a laser as a heat source is that heating can be localized in time and space. By matching the wavelength and beam power to the optical and thermal properties of the material, the amount of heating can be controlled to achieve a specific result, e.g. by choosing a short pulse length heating can be confined to the surface region.

As explained earlier, in order to have a lasting effect on a surface laser radiation must be absorbed. The effect on a given surface, assuming wavelength and beam size are fixed, is determined by the beam power (the rate at which energy is delivered to the surface). A certain proportion of the energy in an incident laser beam, determined by the absorptivity of the surface at that wavelength, is absorbed by the surface and converted to thermal or chemical (if a covalently-bonded material is irradiated by far ultraviolet radiation) energy. The beam power must be high enough so that the energy absorbed is sufficient to bring about a change to the material. If the beam power is too low there will be no lasting effect on the surface. For a given surface at a given wavelength there is a threshold above which the beam power must be raised to cause a change to the surface. Ablation occurs at some point above this threshold.

Ablation

Ablation: the ejection of material from a solid surface as a result of irradiation by a laser beam.

Laser ablation can proceed *photothermally*, where the photon energy is used to heat the surface, or *photochemically*, where the photon energy is used to break chemical bonds within the surface without heating. Photothermal ablation occurs in the visible and infrared parts of the spectrum (long wavelength, low energy photons) while photochemical ablation, which only occurs with covalently-bonded materials, is more likely to occur in the ultraviolet, particularly in the far ultraviolet (short wavelength, high energy photons). Garrison and Srinivasan (1985) have said that the precise wavelength regime in which the photochemical, thermal, or possibly both processes are operative depends on the characteristics of the material. Since the majority of conservation work involving a laser currently uses the fundamental 1.06 μm output from a Nd:YAG laser, most of this chapter is devoted to photothermal ablation and its effects.

Photothermal ablation: temperature rises induced in an absorbing surface

A laser operating at visible and infrared wavelengths (and also at ultraviolet wavelengths on a non-covalently-bonded material) acts as a heat source. The temperature rise induced in the surface of a material by absorption of laser energy depends on a number of factors including the energy absorbed per unit area per unit time by the surface and the physical and thermal properties of the material: density, ρ, specific heat capacity, c, and thermal conductivity, K. The pulse length is also important. The resulting temperature distribution has been calculated by solving a differential equation for heat flow in a solid (Carslaw and Jaeger, 1959; Ready, 1965). Several assumptions have been made including:

1. there is no phase change, e.g. the surface does not melt;
2. the thermal properties of the material are independent of temperature;
3. negligible energy is lost from the surface by reradiation;
4. the absorbed laser energy is uniform across the irradiated area;
5. the laser energy is absorbed within the surface region of the material, i.e. within a depth of several nanometres.

The approximate temperature rise at the surface of the material at time, t, is then given by:

$$T(t) = \frac{2aI}{K}\sqrt{\frac{kt}{\pi}} \quad (3.1)$$

where a is the absorptivity of the surface (at the wavelength of incident radiation), I is the incident laser beam power density, i.e. (pulse energy) ÷ (pulse length × beam size), measured in W/m^2, and k is the thermal diffusivity. Thermal diffusivity determines how rapidly a material will accept and conduct thermal energy and is defined by:

$$k = \frac{K}{c\rho} \quad (3.2)$$

Equation 3.1 is valid for the duration of the laser pulse. At the surface of the material the temperature rises for the duration of the pulse, while energy is being absorbed. Once the laser pulse has finished the temperature falls as heat is conducted into the bulk. The maximum temperature rise at increasing depth below the surface is less and occurs later. Maximum temperature rises of 130°C at the surface and 40°C at 2 μm below have been calculated for aluminium irradiated by a 20 ns laser pulse with $aI = 2 \times 10^{10}$ W/m^2 (Scruby and Drain, 1990). The corresponding temperature rise would be less in a non-metal, e.g. stone, as absorption takes place just below as well as at the surface, to a depth of microns rather than nanometres. In such a material the temperature decays exponentially from a maximum at the surface, while its time-dependence is the same as that of the energy in the pulse. Due to the assumptions made, equation 3.1 does have its limitations in predicting the temperature rise induced in a surface during laser cleaning, in which the thermal properties of the inhomogeneous materials involved are often unknown. It does, however, serve to reinforce the point that it is the beam parameters and material properties together which determine the effect that absorption of laser radiation has on a surface. In some simple cases equation 3.1 can be used to compare the temperature rise induced in different materials. The maximum temperature rise induced by absorption of a Q-switched pulse in a strongly absorbing layer of carbon (a = 0.9) has been estimated as being five times higher than that induced in a weakly absorbing (a = 0.3) layer of calcium carbonate (Cooper, 1994).

If the pulse length is much longer, e.g. 100 μs for a laser operated in normal-mode, then the maximum temperature rise induced in the surface can be two orders of magnitude less than that induced by a Q-switched pulse of the same energy. The reason for this is that the relatively long duration of the normal-mode pulse allows heat to be conducted away from the surface into the bulk while energy is still being deposited into the surface by the laser pulse. A Q-switched pulse deposits heat energy so quickly that very little is conducted away during the pulse. Thermal diffusion lengths (a measure of how far heat penetrates into a material during the laser pulse) of 0.4 μm and 28 μm have been estimated for a limestone sample irradiated by Q-switched and normal-mode pulses respectively.

Photothermal ablation: the effects of a rapid temperature rise in a surface

As mentioned earlier, the temperature rise induced in the surface of a material and any subsequent effects (at a particular wavelength) depend on the power absorbed per unit area of the surface, that is, on the power density of the laser beam and the absorptivity of the surface at that wavelength. In the low power density regime the only effect is a modest increase in temperature of the surface region. This is accompanied by thermal expansion of the heated material and subsequent generation of stresses and strains within the sample (Scruby and Drain, 1990). These thermoelastic stresses generate elastic waves (ultrasound) which propagate deep within the sample (see Figure 3.3). Longitudinal waves have been detected in marble by irradiating the front face of a thin sample, approximately 2 mm thick, with a Nd:YAG laser and attaching a transducer to the back face (Cooper, 1994). No damage was observed to the marble. The signal obtained was less clear than that obtained from an aluminium sample. This was attributed to scattering and strong attenuation of the acoustic energy at crystal boundaries within the marble sample. Laser-generated

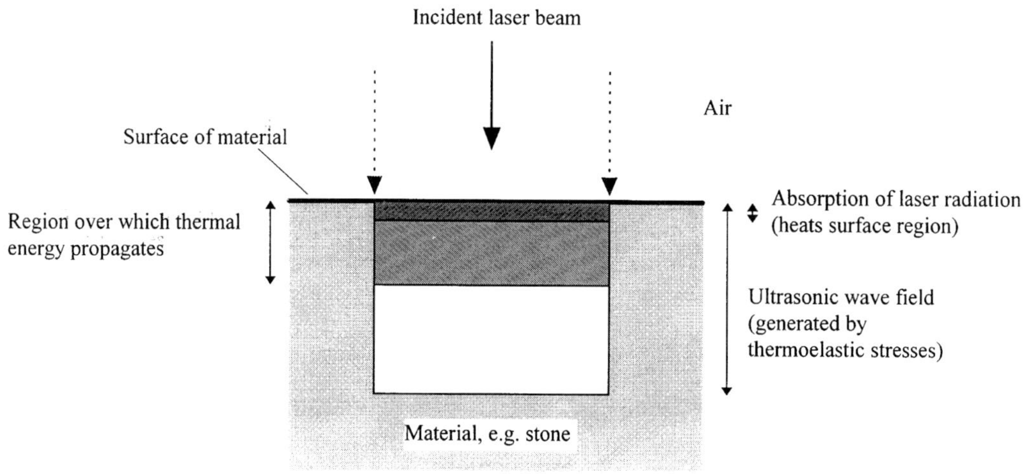

Figure 3.3. *Schematic representation of elastic wave generation within a solid due to absorption of laser radiation.*

ultrasound has been used non-destructively to study anisotropy in rocks (Young and Hutchins, 1987). If the power density is increased, eventually a point is reached where the thermoelastic stress exceeds the yield stress at locations within the material and causes permanent damage. In certain cases the stresses generated are sufficient to eject material from the surface.

If the power density is increased further the temperature rise induced in the surface will eventually lead to phase changes, e.g. melting, vaporization. If the temperature is raised above the melting point the irradiated region of the surface will melt and then resolidify once the laser pulse has finished. If the power density is increased further, vaporization will occur as soon as the boiling point of the surface is exceeded. Vaporization induced by Q-switched laser pulses has been described as a 'thermal explosion' (Ready, 1965); ablation of material away from the surface generates a recoil pressure on the surface, in the same way as a rifle butt moves sharply backwards as the rifle is fired, to conserve momentum.

The different effects caused by absorption of pulsed laser radiation at visible and infrared wavelengths are summarized in Figure 3.4. The effects described are typical of those brought about by absorption of Q-switched pulses where the temperature rise induced at the surface is extremely rapid, typically 10^9–10^{11} °C/s. For normal-mode pulses the rate of temperature rise is several orders of magnitude less and ablation is believed to proceed simply by vaporization.

Photochemical ablation

As explained earlier, photons emitted in the far ultraviolet are sufficiently energetic to break many covalent chemical bonds directly upon absorption without heating. Ultraviolet radiation initiates a chemical reaction in which the products have a larger volume than the original sample. The products are ablated from the surface as a result of this sudden volume increase (Garrison and Srinivasan, 1985). Photochemical ablation is able to remove small amounts of material in a very precise manner. This is dealt with in more detail in Chapter 6.

Summary

The paths by which laser ablation occurs are summarized diagrammatically in Figure 3.5. For our purpose it is sufficient to assume that photothermal ablation is the dominant process for laser cleaning at infrared wavelengths, and photochemical ablation is the dominant

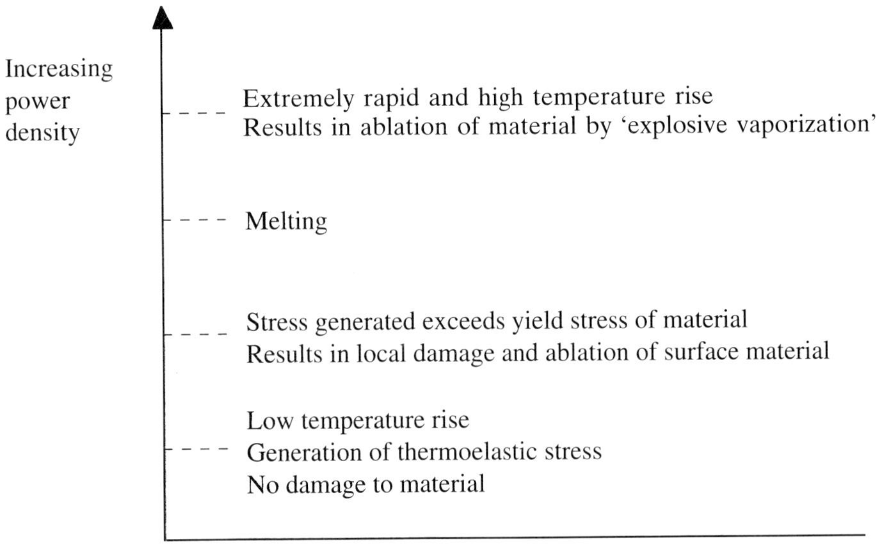

Figure 3.4. *Summary of the typical effects caused by absorption of Q-switched laser radiation (visible and infrared wavelengths).*

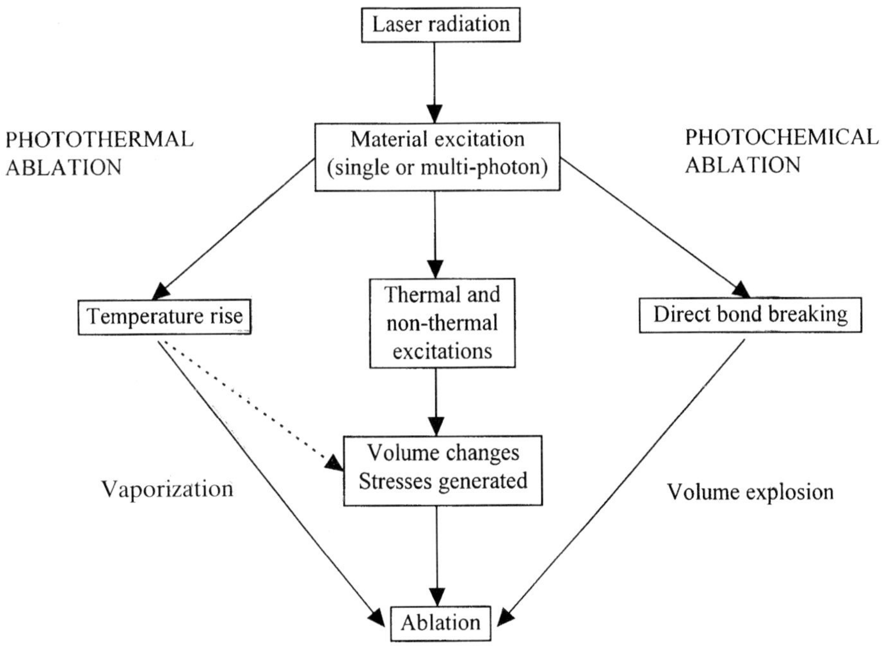

Figure 3.5. *Different stages of pulsed laser ablation. The left side represents photothermal ablation while the right side represents photochemical ablation. The central path describes a combination of both processes.*

process at ultraviolet wavelengths. A path has been included for the case of ablation by a combination of photothermal and photochemical processes which can occur at intermediate wavelengths.

Other effects

Other effects brought about by absorption of pulsed laser radiation are briefly described in this section.

Plasma formation

Plasma formation occurs when material is ablated from a surface at very high power densities (Bauerle, 1996). Irradiation of the surface leads to ejection of material very early during the laser pulse. The plume of vapour formed just above the surface absorbs a fraction of the energy of the incoming laser pulse. This gives rise to intense heating and ionization of the ejected material generating an extremely intense spark, or plasma. The formation of a plasma next to the surface dominates the later transfer of energy and momentum to the surface. The plasma strongly absorbs energy from the laser pulse and becomes extremely hot. Once the particle density in the plasma reaches a critical value the plasma acts as a shield, preventing the energy in the laser pulse from reaching the surface. At this point the energy is strongly absorbed in a very thin layer of the plasma, which is heated intensely, expands and produces an impulse reaction on the surface. Once the laser pulse has finished the plasma expands away from the surface and dissipates.

Acoustic pulse formation

At low power densities (no ablation), absorption of laser radiation and subsequent heating and thermal expansion of a surface leads to rapid expansion and compression of the air molecules immediately above the surface. At higher power densities particle ablation, which occurs at velocities of the order 100 m/s, is sufficiently rapid to generate an acoustic wave in the air above the irradiated region which is audible as a 'snapping' sound (Tam, Leung, Zapka and Ziemlich, 1992). At extremely high power densities the plasma contributes to the generation of a shock pulse. The amplitude of the acoustic wave generated in air as a result of absorption of laser radiation is dependent on the intensity of the interaction between the laser pulse and surface. It has been suggested that this could be used by conservators during cleaning to establish power densities at which dirt removal can be accomplished without damage to the surface of the object (Cooper, Emmony and Larson, 1995).

Radiation pressure

Laser radiation incident on a solid surface generates a pressure due to the change in momentum of the photons as they are reflected or absorbed at the surface (Scruby and Drain, 1990). The forces and stresses on a solid resulting from radiation pressure are generally several orders of magnitude less than those due to thermoelastic effects and ablation, and can be discounted when considering cleaning mechanisms.

Summary

The interaction of laser radiation with a solid surface is a complex process. There are many effects caused by absorption of laser radiation and these depend on the chemical and physical properties of the surface as well as the beam parameters. These effects are summarized in Figure 3.6. In the high power density regime all of these phenomena apply. At low power densities only some apply.

Laser cleaning in industry

Much of the research carried out into laser cleaning has been concerned with the removal of contaminants from substrates used in the semiconductor industry. This interest arose from a dissatisfaction with other cleaning techniques, such as ultrasonics and high pressure jets, which worked poorly for submicron particles and damaged some delicate substrates. The situation was analogous to concern over the effects of traditional methods

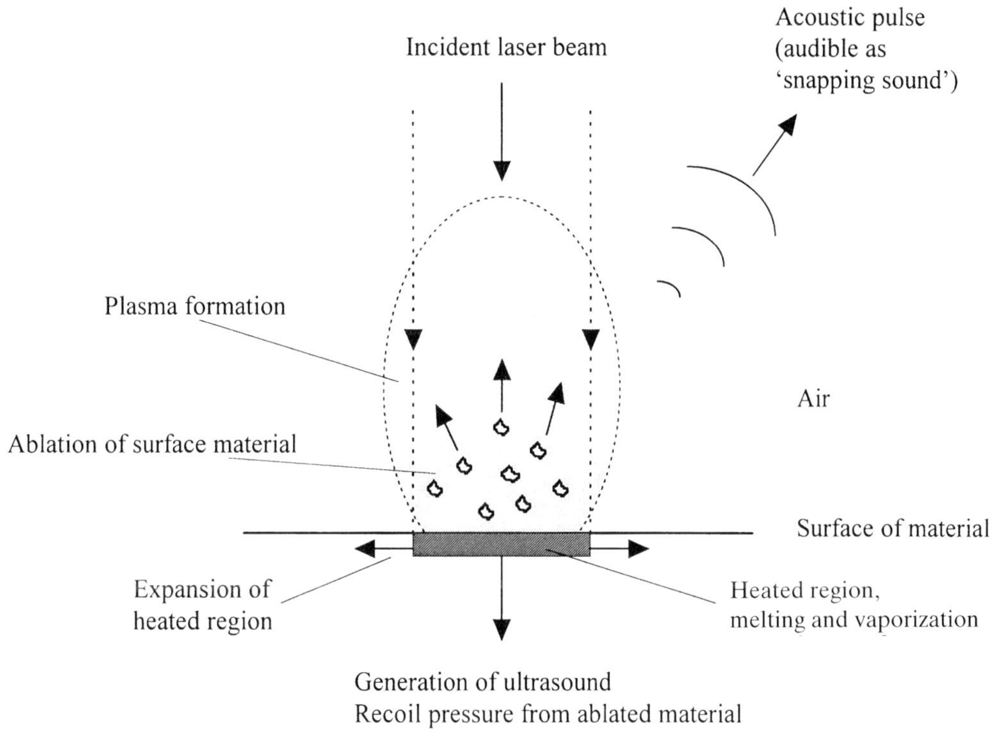

Figure 3.6. *Schematic representation of the effects of absorption of a high power laser beam.*

of cleaning employed by conservators on delicate art objects. In both cases there was a need for a more effective and selective method of removing strongly-adhered contaminants without damaging a delicate substrate surface. A brief review of the published literature is presented here. Previous work in this field is used later in this chapter to help explain some of the mechanisms behind laser cleaning in conservation.

Initial work in the semiconductor industry

During the 1990s many papers have been published in the area of laser removal of micron-sized particles from surfaces. In 1991 Kelley used a Q-switched Nd:YAG laser (pulse length 20 ns) to remove micron-sized tungsten particles from a lithium niobate substrate. Working at an energy density of 0.65 J/cm^2, 95% of the particles were removed after 100 pulses. Kelley proposed a model to describe the process based on the idea that strong absorption of laser energy by a particle gives rise to a rapid thermal expansion which generates forces and accelerations large enough to eject the particle from the surface (Kelley, 1991). In the same year Imen and co-workers published research in which they achieved Laser-Assisted Particle Removal (LAPR) by applying a layer of water to the surface and irradiating with pulses from a CO_2 laser (Imen, Lee and Allen, 1991). Alumina particles were removed from a silicon substrate without any damage to the substrate using pulses of 1 μs duration and an energy density of 30 J/cm^2. Particles as small as 1 μm were removed. The authors described the process qualitatively. Water is preferentially absorbed in the capillary spaces under and around the particles. At the CO_2 wavelength the beam energy is mainly absorbed by the water molecules which evaporate explosively. The forces generated are much greater than the adhesion forces between the particle and substrate and propel the particles away from the surface. Irradiation of the surface without a water film did not result in particle removal. The authors concluded that the LAPR technique 'offers a unique solution with a high probability of success for small particle removal in the microelectronics and optics industries'.

The effect of wavelength on cleaning

Researchers soon began to experiment with different wavelengths of laser radiation and different liquid films. In 1991 Zapka and co-workers published research in which pulsed ultraviolet and infrared laser radiation were successfully used to remove submicron-sized particles from delicate silicon wafer surfaces and lithography membrane masks (Zapka, Ziemlich and Tam, 1991). Effective removal of 0.35 μm alumina particles was achieved using a KrF excimer laser at a wavelength of 248 nm (pulse length 16 ns) with an energy density of 0.35 J/cm^2. The authors observed that: at safe (for the substrate) energy densities particles are ejected intact and at high speed approximately perpendicular to the surface (particle removal does not occur by melting and vaporization); removal occurs only in the irradiated area; efficient removal occurs above a certain threshold energy density which depends on the size and type of the particle and on the type and finish of the substrate. The underlying mechanisms were believed to be a combination of those suggested by Kelley and Imen:

1. strong acceleration of the particle by an extremely rapid expansion of either the substrate or the particle itself (depending on the wavelength of radiation chosen).
2. explosive boiling of a thin water film (condensed from the atmosphere) between the particle and substrate as a result of strong energy transfer from the laser beam.

The effect of water on laser cleaning efficiency

Zapka and co-workers also discovered that the removal efficiency was enhanced by applying a water film approximately 1 μm thick to the surface just prior to irradiation (Zapka, Ziemlich and Tam, 1991). Alumina particles were removed effectively by both ultraviolet and infrared radiation. Gold particles as small as 0.2 μm were completely removed from a substrate by pulsed ultraviolet radiation, but were only partially removed by pulses from a Q-switched erbium:YAG (Er:YAG) laser, selected for its strong absorption in water, operating at 2.94 μm. From this it was concluded that 'for effective particle removal, it is generally preferable to choose the incident laser wavelength to be strongly absorbed by the substrate, rather than by the thin water film' so that heating of the water film occurs at the water/substrate interface. The authors agreed with Imen and co-workers that the removal of particulate and film contaminants by pulsed laser radiation was a promising technique for ultraclean manufacturing processes.

In 1992 Tam and co-workers published results showing successful removal of gold, alumina, silicon and latex particles, ranging in size from 10 μm down to 0.1 μm, from various surfaces including silicon wafers and gold (Tam, Leung, Zapka and Ziemlich, 1992). By this stage several varieties of LAPR had been developed: laser cleaning could be carried out with or without a simultaneous deposition of a liquid film; wavelengths could be chosen so that absorption of energy occurs mainly at the sample surface, in the liquid, in the particulate or some combination of the three.

In their 1992 publication Tam and co-workers discussed each of these techniques. Cleaning without a liquid film was termed *dry* laser cleaning. In the case where absorption of energy occurs mainly in the substrate Tam and co-workers estimated that although the sudden expansion induced in the substrate surface might only be several nanometres, the acceleration generated would be of the order 10^8 m/s^2, due to the short timescale involved (10^{-8} s). This is far in excess of gravitational acceleration and sufficient to 'shake' particles from the surface. In cases where there is strong absorption by the particle Tam and co-workers were in agreement with Kelley: a similar acceleration applies to the particle, again leading to ejection from the surface. The authors also stated that strong particle absorption is a very selective technique, useful for removal of specific particles (strong absorption) from a specific substrate (weak absorption). This is in fact the 'variety' of laser cleaning most often employed by the conservator. Highly selective cleaning of many artworks is possible with a Nd:YAG laser since the particulates which make up the soiling are strongly absorbing at 1.06 μm but the delicate substrate is only weakly absorbing.

Cleaning with a liquid film was termed *steam* laser cleaning. Tam and co-workers noted that fast ablation of the liquid film, which produces the large transient forces to overcome the particle/substrate adhesion, is caused by three types of heating: substrate heating only, liquid film heating only, and both substrate and liquid film heating. The highest laser cleaning efficiency is provided by strong substrate heating. In cases where a water film is used with laser cleaning in conservation absorption of the energy occurs mainly in the particulates which make up the soiling. Ablation of the liquid film then proceeds in a similar way to the 'strong substrate heating' case. Tam and co-workers suggested that the enhanced cleaning effect (more particles of smaller size removed at lower energy density with fewer pulses) provided by a liquid film is important in cases where the energy density and number of pulses must be kept to a minimum to avoid substrate damage. The authors estimated that laser cleaning by strong substrate absorption can lead to a peak temperature rise of 370°C in some substrates. Steam laser cleaning has been performed on delicate objects, including a 3 μm thick silicon membrane coated surface, without any detectable surface damage. This is thought to be due to the peak temperature existing for only a very short time. The authors warn however that a quantitative theoretical understanding of liquid film enhanced laser cleaning (temperatures and pressures generated at the surface, particle ejection velocities, etc.) is lacking.

In conservation, laser ablation is usually achieved by strong particle absorption for two main reasons: (1) surface contaminants must be removed in such a way that the peak temperature rise and pressure generated at the object surface, and hence the probability of damage, are minimized, and (2) the density of contaminants is usually such that the laser pulse cannot reach the object surface. Many contaminants encountered in conservation absorb strongly in the near infrared. In cases where the underlying object surface absorbs weakly in the near infrared the Nd:YAG laser can be used to clean selectively.

Other industrial applications of laser cleaning

During the early 1990s research into cleaning by laser ablation was undertaken in many different areas. These included: cleaning of glass by CO_2 and nitrogen lasers (Ueda, Makino and Kagawa, 1991), removal of fingerprints from glass and quartz surfaces by KrF excimer laser (Lu, Komuro and Aoyagi, 1994) and stripping of polyurethane-coated wires by KrF excimer and CO_2 lasers (Brannon, Tam and Kurth, 1991). Much of the work carried out in laser cleaning has used the laser as a source of heat for generating laser ablation. Research into laser ablation of organic polymers, however, has shown that in the deep ultraviolet ablation proceeds by photochemical ablation (Garrison and Srinivasan, 1985; Srinivasan and Braren, 1989). Garrison and Srinivasan presented models to describe the differences between photothermal and photochemical ablation. The photochemical model predicted that etching of an organic polymer by far ultraviolet laser radiation, e.g. 193 nm, would occur cleanly, leaving behind a well-defined pit. The photothermal model predicted that etching of the same polymer by longer wavelength radiation, e.g. 532 nm, would leave the sample near the pit distorted and melted. This was found to be the case. The discovery that far ultraviolet laser radiation can cleanly ablate polymeric and biological materials has led to the use of excimer lasers in many diverse fields, including eye surgery (removal of corneal tissue) and photoresist etching. In conservation the controlled layer-by-layer removal offered by photochemical ablation has found applications in the cleaning of paintings and stained glass where photothermal ablation cannot proceed selectively due to the widely differing absorption properties of the materials present.

Research undertaken in the field of laser cleaning has led to a number of commercial industrial processes, in both small and large scale cleaning. A company in the United States has patented a technique in which far ultraviolet laser radiation in combination with a flowing inert gas is used in the semiconductor industry to remove microcontaminants from wafers, photomasks, flat panel display substrates and processing chamber interiors (Engelsberg, 1995). The flow of gas across the surface being cleaned prevents reattachment of the contaminant. The process is able to remove particles from 80 μm to below 0.1 μm in size, metallic ions, organic films and oxides without leaving any residues or causing detectable surface damage. The process is claimed to be simple, flexible, non-destructive, economical and environmentally benign.

At the opposite end of the scale, a German manufacturer has developed a short pulsed CO_2 laser capable of generating 2 kW average power for paint-stripping of aircraft (Anonymous, 1995). Pulses of far infrared radiation are able to vaporize paint layers without thermal damage to the underlying metallic or fibre-reinforced plastic surface. Air extraction and filters are used to remove the waste products, which are claimed to amount to just one-fiftieth of the waste created by chemical techniques. The company claim that the surface of a wide-body aircraft (about 1600 m^2) can be stripped in three days using three lasers working in parallel.

Studies of laser ablation in conservation

Within conservation there have been no detailed studies of the mechanisms involved in the laser cleaning process. This is probably due to two main reasons: interest in the field has only become focused relatively recently, and it is an extremely diverse field involving a complex combination of laser parameters and surface conditions. Laser cleaning in conservation involves laser interactions with an incredibly wide range of contaminants, such as pollution encrustations, corrosion layers, organic growths and previous treatments, on an equally wide range of substrates including many different types of stone, metal, ceramic and paper. Since art objects cannot be tested destructively (samples for testing are often very difficult to obtain) and the surface conditions are usually extremely difficult to simulate (some of the pollution encrustations and corrosion layers have formed over hundreds of years in an often unique and evolving atmosphere), most researchers have concentrated on the quality of cleaning achieved using the laser and not on the mechanisms by which it has been achieved.

As lasers become more established in the conservation world, understanding of the mechanisms involved in the cleaning process will undoubtedly improve as more research groups become involved in the search for solutions to specific problems and in development of improved and more versatile laser cleaning systems. Asmus and co-workers presented a model which provided a qualitative understanding of the different mechanisms involved in cleaning by long-pulse and Q-switched laser radiation (see Chapter 1). This model, although useful at the time, failed to take account of the different mechanisms (some more selective than others) by which cleaning can be achieved in Q-switched mode and their relative importance depending on factors such as the incident energy density. This chapter provides a qualitative description of the main processes involved in laser cleaning, drawn from the author's own experience and the published literature in this field. The interested reader who would like to gain a more thorough understanding of laser ablation is referred to books by Miller (1994b), von Allmen and Blatter (1995) and Bauerle (1996).

The explanations presented in this chapter will hopefully provide the conservator with an improved understanding of laser cleaning, allowing them to make sensible decisions on when and how to use the technique. As explained in Chapter 2, the most commonly used laser in conservation studios is currently the Nd:YAG which is reliable, rugged and provides radiation at a wavelength which is selectively absorbed by many dirt layers on a wide range of materials (see Chapter 5). The remainder of this chapter has, therefore, been devoted to the mechanisms by which cleaning is believed to occur at near infrared wavelengths. Processes occurring at ultraviolet wavelengths are dealt with in Chapter 6.

Mechanisms involved in near infrared laser cleaning in conservation

The mechanisms involved in laser cleaning of artworks are not fully understood. To the knowledge of the author no research has yet been published in which the temperatures and pressures generated at the surface during the process have successfully been measured. To date research has been carried out very much on a qualitative basis. During the 1970s Asmus and co-workers proposed two main mechanisms to describe laser cleaning at visible and infrared wavelengths: a selective vaporization model for cleaning by normal-mode laser pulses and a plasma-generated shock wave model for cleaning by Q-switched pulses. These models are described in Chapter 1. Asmus felt that Q-switched pulses were more efficient in removing dirt from surfaces but less selective, i.e., cleaning by Q-switched laser radiation is more likely to cause damage to the artefact. Research since has found that, in general, provided cleaning by Q-switched pulses is carried out at an energy density below that necessary for plasma formation then the process is not only more efficient but also more selective than cleaning by normal-mode pulses (Orial and Gaufillet, 1989; Cooper, Emmony and Larson, 1992; Cooper, 1994; Larson and Cooper, 1996).

In normal-mode, cleaning is believed to proceed via the 'selective vaporization' model proposed by Asmus. However, conduction of thermal energy into the bulk of the material is significant due to the relatively long pulse length. The temperature rise induced in the surface region of the absorbing material is, therefore, usually too low and too slow for efficient ablation. Melting of the underlying object surface has been reported (Cooper, 1994). Figure 3.7 shows an electron micrograph of a ceramic surface cleaned by normal-mode pulses. The crazing pattern has been attributed to thermally-induced cracking in the surface which has also revitrified forming a

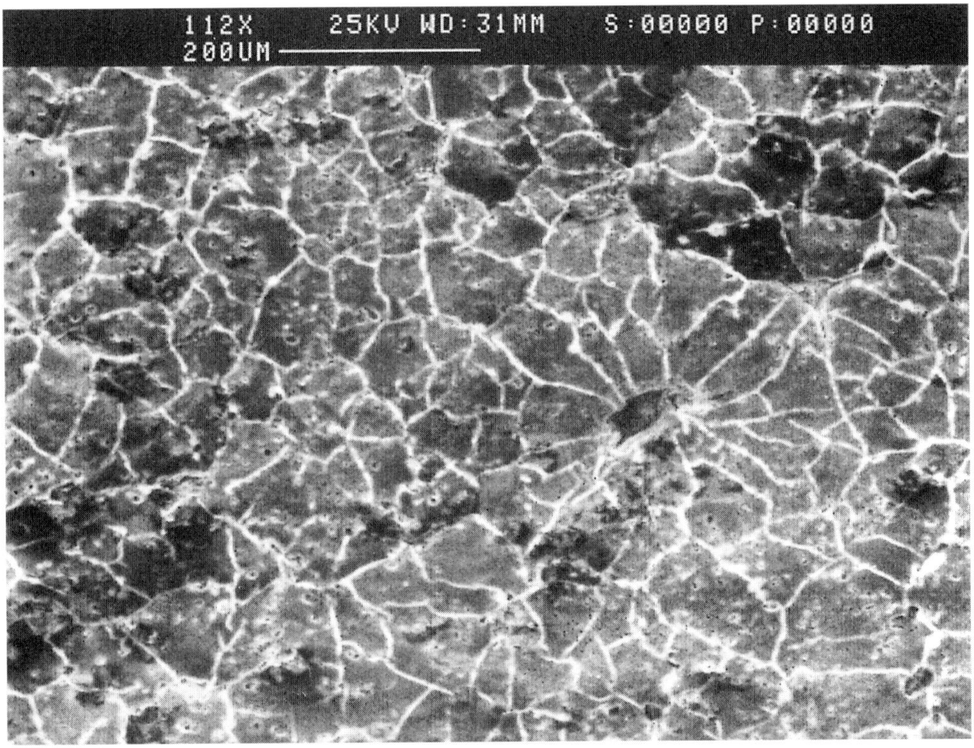

Figure 3.7. *Thermally-induced cracking in the surface of a ceramic tile cleaned by normal-mode Nd:YAG laser radiation.*

transparent glassy layer (Larson and Cooper, 1996). In Q-switched mode the rate of energy deposition into the absorbing surface is so fast that thermal conduction into the underlying material is negligible. The subsequent ablation of surface material is believed to occur via a process involving any one, or a combination, of four main mechanisms. The dominant mechanism and hence the selectivity of the process is determined by the incident energy density, the type of dirt and its bonding to the surface. Since the fundamental output of the Nd:YAG laser often used in conservation is in the infrared part of the electromagnetic spectrum, ablation of absorbing material occurs photothermally. This section provides a qualitative description of the main mechanisms by which Q-switched Nd:YAG laser cleaning (at 1.06 μm) is believed to occur, based on observations made during cleaning, detailed examinations of cleaned surfaces and a study of the published literature, much of which is described earlier in this chapter.

Rapid thermal expansion

At relatively low energy density the most important mechanism is believed to be rapid thermal expansion as described by Kelley (1991) and Tam and co-workers (1992). Consider a single particle of dirt strongly adhered to the surface of an artefact (see Figure 3.8). Strong absorption of the laser pulse by the particle leads to an extremely rapid and large temperature rise. The heated region undergoes a small but rapid thermal expansion generating forces sufficient to overcome the particle's adhesion to the surface. The resultant force is away from the surface and leads to particle ejection. Figure 3.9 shows ejection of particulate matter from a terracotta surface. In situations where the dirt absorbs strongly and the underlying surface absorbs weakly, this process is extremely selective and cleaning, even on delicate surfaces, can be undertaken without causing damage. This process applies equally to the removal of encrustations from a surface; generation of thermoelastic stresses within the irradiated region is usually sufficient to break bonds within the crust and also between the crust and the underlying surface. Additional energy absorbed from the laser pulse is used as kinetic energy in ejection of solid particles from the surface. Evidence for this mechanism operating at low energy density is that: (1) there is no visible plasma generation, and (2) the ejected material appears to consist mainly of particulate matter. As described earlier in this chapter, cleaning efficiency (rate of cleaning and degree of cleaning) can sometimes be enhanced by applying a thin film of water to the dirt surface immediately prior to irradiation by the laser. The water is adsorbed onto the dirt particles (see Figure 3.10) and penetrates into the pores

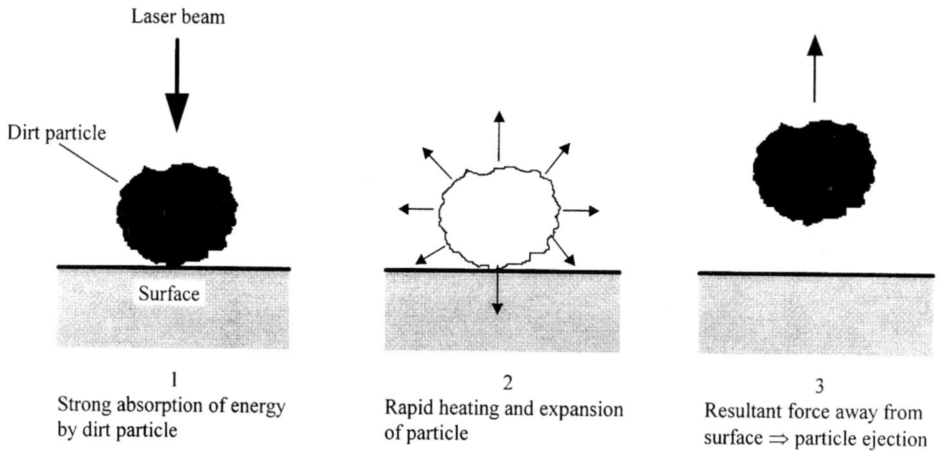

Figure 3.8. *Removal of dirt particles by rapid thermal expansion.*

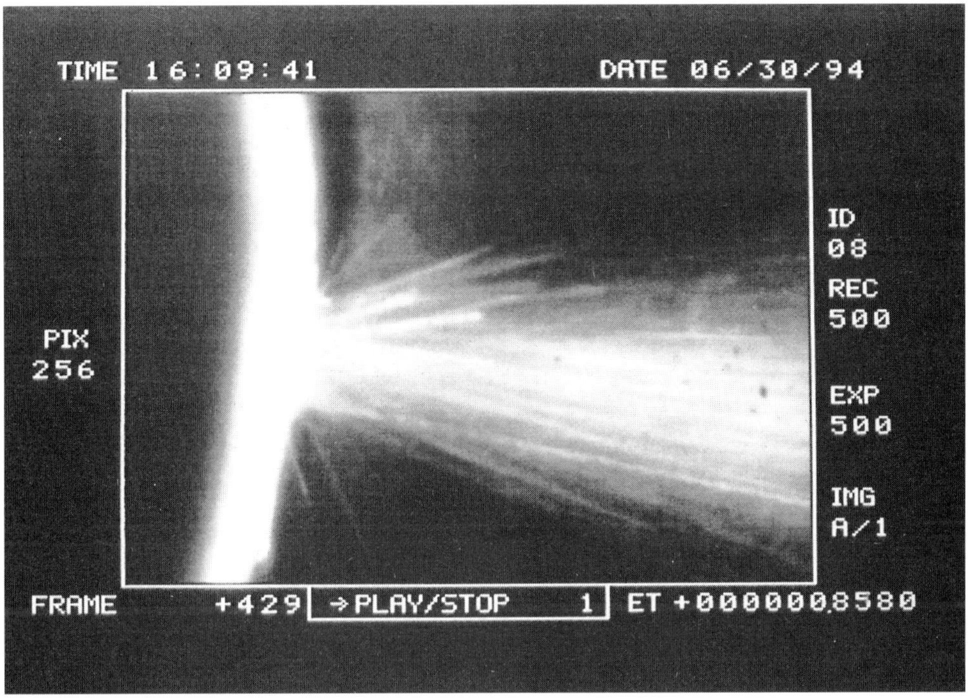

Figure 3.9. *Ejection of particulate matter from a polluted terracotta surface irradiated by Q-switched Nd:YAG laser radiation (photographed by high speed camera). The surface is viewed side-on with the laser beam incident from the right.*

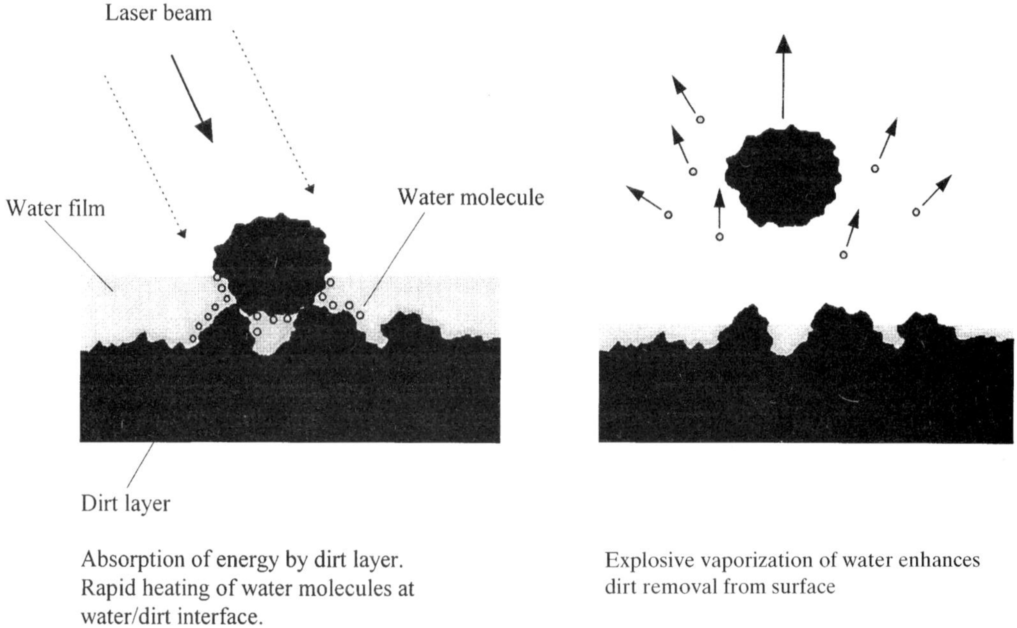

Figure 3.10. *Steam laser cleaning.*

and cracks within the dirt layer. The dirt layer absorbs energy from the laser pulse as normal and rapid heating of the liquid film at the dirt/water boundary leads to explosive vaporization of the water molecules. The forces exerted on and within the dirt layer during this process are sufficient to eject further material from the surface. The same effect does not happen on a clean surface since the water film is not heated to the same extent due to weak absorption of energy by the surface.

Explosive vaporization

At slightly higher energy densities some material is heated sufficiently to bring about vaporization. Since energy is deposited into the surface extremely quickly the rate of heating is very fast and vaporization takes place explosively (Ready, 1965). The effects of this explosive vaporization on the substrate can be difficult to control and the mechanism should therefore be avoided if possible, i.e. if a more selective mechanism is effective.

Surface relaxation due to plasma effects

At high energy density, sometimes one order of magnitude greater than that needed for ablation by rapid thermal expansion, surface relaxation is thought to become the dominant process. This is the mechanism proposed by Asmus in 1973 (see Figure 1.4). Material vaporized very early during the laser pulse leads to the creation of a plasma just above the surface. Rapid expansion of the plasma applies a compression to the irradiated region of the surface. Once the laser pulse has finished the surface relaxes. This very fast compression and relaxation of the surface can lead to ejection of more material from the surface. This mechanism is believed to be the least selective of those mentioned since the forces generated by the expanding plasma cannot discriminate between the polluted and clean surfaces.

Acoustic shock-induced surface disruption

Acoustic shock-induced surface disruption has been suggested as another mechanism by which material can be removed at high energy density (Cooper, 1994). Propagation of an acoustic shock pulse along the surface of a crust can generate stresses sufficient to cause physical disruption of material in the vicinity of the irradiated region. Unfortunately, it can also cause disruption to weakened material in the surface of the object, e.g. if dirt is lodged in a surface crack.

Summary

The interaction of a Q-switched laser pulse with a contaminated surface is a complex process. The process becomes even more complex as the incident energy density is increased. At low energy density removal of material occurs mainly via *rapid thermal expansion* which is thought to be the most selective mechanism. At high energy density the less selective mechanisms of *surface relaxation* and *acoustic shock-induced surface disruption* become dominant. At intermediate energy densities *explosive vaporization* (without plasma formation) plays an important role. The mechanisms by which laser cleaning of artworks occurs have not been studied in detail, but those described here are believed to be the most likely. The precise mechanisms, or combination of mechanisms, operating at a particular time depend on many factors including the incident energy density, the type of contaminant and the strength of its bonding to the underlying surface. It is the job of the conservator to carry out cleaning in such a way that dirt removal is achieved by the most selective mechanism possible, thereby minimizing the risk of object damage.

The self-limiting nature of laser cleaning

Laser cleaning at 1.06 μm is often described as a *self-limiting* process. In this respect self-limiting means that ablation of material from the surface of an object stops as soon as the dirt layer has been removed, without any interference from the conservator, i.e. the clean surface is not damaged by exposure to further laser pulses. The ideal case for laser cleaning is a perfectly absorbing (a = 1) dirt layer on a perfectly reflecting (a = 0) substrate. Such a case is obviously self-limiting since the energy in the laser pulse is completely reflected from the substrate once the dirt has been removed. A laser beam can have no effect on a surface unless some of its energy is absorbed. In reality all materials absorb at least a small fraction of energy at 1.06 μm. By increasing the incident energy density a point is eventually reached for all materials where sufficient energy is absorbed to cause some form of damage to the surface. Fortunately in many cases at 1.06 μm the dirt layer has a higher absorptivity than the substrate.

The relatively high absorptivity of a typical dirt layer and its other material properties, such as specific heat capacity, density and thermal conductivity, often combine to give a relatively low ablation threshold. The *ablation threshold* is the energy density above which ablation occurs. If the ablation threshold of the dirt layer is less than that of the substrate, i.e. the dirt is more easily removed than the substrate is damaged, then self-limiting cleaning is feasible. Working within this regime allows the conservator to clean a surface without 'overcleaning', a major drawback of conventional cleaning methods: the conservator is able to go back over an area to remove remnants of dirt without damaging surrounding areas which have already been cleaned. If the ablation threshold of the dirt layer is equal to or greater than that of the substrate then the process will not be self-limiting. This situation is similar to that of conventional methods of cleaning, such as air-abrasive, where the conservator must avoid exposing the cleaned surface to further pulses.

Self-limiting cleaning has been observed in many cases, e.g. removal of strongly absorbing pollution encrustations from weakly absorbing marble and limestone sculpture (Cooper, 1994). This is demonstrated graphically in Figure 3.11. If cleaning is carried out at an energy density below E_c but above E_p then the process will be self-limiting. The slope of the plots is a result of the inhomogeneous nature

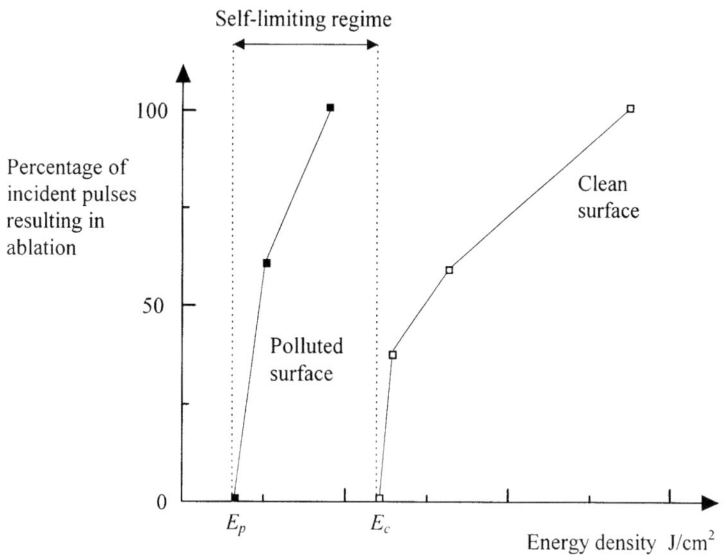

E_p: minimum energy density required to remove pollutant.
E_c: energy density above which damage to clean surface can occur.

Figure 3.11. *Graph showing ablation thresholds (Q-switched Nd:YAG) for clean and polluted Lincoln limestone surfaces. The region between E_p and E_c is self-limiting (Cooper, 1994).*

of the surfaces, i.e. some areas of a surface are more easily ablated than others. It is important to realise that since the ablation threshold depends to some extent on the condition of the surface, (it is easier to remove material from a badly decayed friable surface than it is from a surface in sound condition), cleaning is much more likely to be self-limiting if carried out at low energy density so that the most selective mechanism operates.

References

Anonymous (1995) CO_2 lasers reduce waste from aircraft paint removal. *Opto and Laser Europe*, **18** March, 42–43

Bauerle, D. (1996) *Laser Processing and Chemistry*, Springer-Verlag

Brannon, J.H., Tam, A.C. and Kurth, R.H. (1991) Pulsed laser stripping of polyurethane-coated wires: a comparison of KrF and CO_2 lasers. *Journal of Applied Physics*, **70** (7), 3881–86

Breech, F. and Cross, L. (1962) *Applied Spectroscopy*, **16**, 59

Carslaw, H. and Jaeger, J. (1959) *Conduction of Heat in Solids*, Oxford University Press

Cooper, M.I. (1994) *Laser Cleaning of Stone Sculpture*. PhD Thesis, Loughborough University

Cooper, M.I., Emmony, D.C. and Larson, J.H. (1992) A comparative study of the laser cleaning of limestone. *Proceedings of the 7th International Congress on Deterioration and Conservation of Stone* (J. Delgado Rodrigues, F. Henriques and F. Telmo Jeremias, eds.), Lisbon, June 1992, 1307–15

Cooper, M.I., Emmony, D.C. and Larson, J.H. (1995) Characterization of laser cleaning of limestone. *Optics and Laser Technology*, **27** (1), 69–73

Engelsberg, A.C. (1995) Laser-assisted cleaning proves promising. *Precision Cleaning*, May

Gagliano, F.P. and Zaleckas, V.J. (1972) Laser processing fundamentals. In *Lasers in Industry* (S.S. Charschan, ed.) pp. 150–76, Van Nostrand Reinhold Company

Garrison, B.J. and Srinivasan, R. (1985) Laser ablation of organic polymers: microscopic models for photochemical and thermal processes. *Journal of Applied Physics*, **57** (8), 2909–14

Hecht, E. (1987) *Optics* 2nd Edition, Addison-Wesley Publishing Company

Imen, K., Lee, S.J. and Allen, S.D. (1991) Laser-assisted micron scale particle removal. *Applied Physics Letters*, **58** (2), 203–205

Kelley, J.D. (1991) Particle removal from surfaces by pulsed laser irradiation. *SPIE Laser Induced Damage in Optical Materials*, **1624**, 153–60

Larson, J.H. and Cooper, M.I. (1996) The use of laser energy for cleaning architectural terracotta decoration. *Architectural Ceramics; Their History, Manufacture and Conservation*, Joint Symposium of English Heritage and UKIC, September 1994, 92–100, James & James

Lu, Y.F., Komuro, S. and Aoyagi, Y. (1994) Laser-induced removal of fingerprints from glass and quartz surfaces. *Japanese Journal of Applied Physics*, **33** (1), 4691–96

Miller, J.C. (1994a) History, scope and the future of laser ablation. In *Laser Ablation: Principles and Applications* (J.C. Miller, ed.) pp. 1–9, Springer-Verlag

Miller, J.C. (1994b) *Laser Ablation: Principles and Applications*, Springer-Verlag

Orial, G. and Gaufillet, J.P. (1989) Nettoyage des monuments historiques par désincrustation photonique des salissures. *Technologie Industrielle Conservation Restauration du Patrimonie Culturel*, Colloque AFTPV/SFIIC, Nice, 19–22 September, 118–25

Ready, J.F. (1965) Effects due to absorption of laser radiation. *Journal of Applied Physics*, **36** (2), 462–68

Scruby, C.B. and Drain, L.E. (1990) *Laser Ultrasonics*. Adam Hilger

Srinivasan, R. and Braren, B. (1989) Ultraviolet laser ablation of organic polymers. *Chemical Reviews*, **89**, 1303–16

Tam, A.C., Leung, W.P., Zapka, W. and Ziemlich, W. (1992) Laser cleaning techniques for removal of surface particulates. *Journal of Applied Physics*, **71** (7), 3515–23

Ueda, M., Makino, R. and Kagawa, K. (1991) Laser cleaning of glass. *Optics and Lasers in Engineering*, **15**, 275–78

Von Allmen, M. and Blatter, A. (1995) *Laser-Beam Interactions with Materials: Physical Principles and Applications*, Springer-Verlag

Young, R.P. and Hutchins, D.A. (1987) Measuring anisotropy in rocks using laser-generated ultrasound. *Geophysics Journal of the Royal Astronomical Society*, **91**, 501–16

Zapka, W., Ziemlich, W. and Tam, A.C. (1991) Efficient pulsed laser removal of 0.2 μm sized particles from a solid surface. *Applied Physics Letters*, **58** (20), 2217–19

4

Practical laser cleaning with a Q-switched Nd:YAG laser

Martin Cooper

This chapter describes some of the practical aspects of laser cleaning. It looks at the most common type of laser used in conservation studios and describes some of the practical techniques employed by conservators to optimize the quality of cleaning achieved using this laser. If used appropriately and skilfully laser cleaning at near infrared wavelengths is an extremely selective process on a wide range of materials, suited to removing dirt layers from one tenth of a millimetre to several millimetres thick. As a result of its selectivity, the process has been incorporated into a handheld tool. Laser cleaning at ultraviolet wavelengths is suited to removing unwanted layers in an even more precise manner: layers as thin as 0.1 μm can be removed. This layer-by-layer process can be applied to removing surface layers of up to about 300 μm thickness. Although more precise than cleaning at near infrared wavelengths, cleaning in the ultraviolet does not possess the same degree of selectivity, and on-line monitoring and diagnostic techniques must, therefore, be used to ensure that the substrate is not damaged during cleaning. The necessity for additional techniques means that the laser must be incorporated into a workstation rather than simply developed into a handheld tool. A handheld ultraviolet cleaning tool may find applications in the future where cleaning at other wavelengths is ineffective, such as the removal of certain organic-based substances (e.g. algae, lichens). The practical aspects of cleaning with ultraviolet lasers are discussed in Chapter 6. The advantages and disadvantages of laser cleaning, and its health and safety implications and training requirements are also discussed in this chapter.

A laser cleaning tool

As mentioned in earlier chapters the most commonly used cleaning laser in conservation studios is the Q-switched Nd:YAG laser, emitting laser radiation at a wavelength of 1.06 μm (near infrared) in short pulses of typically 5–10 ns duration. Since the mid-1960s much work has gone into development of the Nd:YAG laser which has now become a very reliable, easy to maintain, compact and rugged laser. These are essential characteristics for any cleaning tool. Most importantly, however, the Nd:YAG laser emits the right 'type' of laser radiation for selective cleaning of a wide range of surfaces. Although developed during the late 1980s as a tool primarily for selective removal of black pollution encrustations from marble and limestone sculpture, the Nd:YAG laser has since been used successfully to clean many other surfaces including sandstone, terracotta, plaster, bone, ivory and parchment (see Chapter 5). It is an extremely versatile tool. Near infrared radiation is absorbed strongly by many types of soiling but only

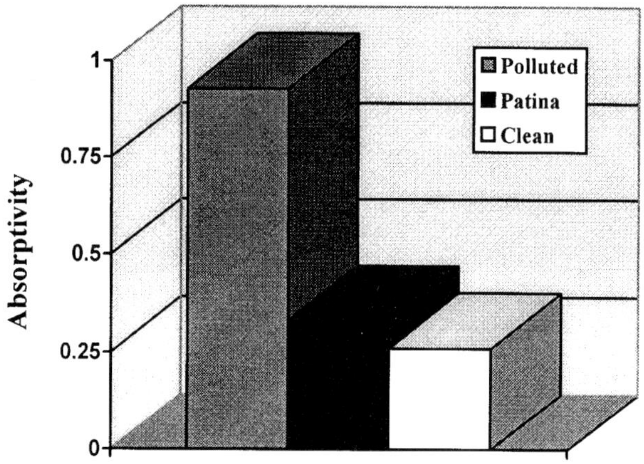

Figure 4.1. *Absorptivity of clean and polluted surfaces of a Carrara marble sculpture, measured at 1.06 μm.*

weakly by the surfaces of many artworks. The short pulse length of the Q-switched Nd:YAG laser ensures that effective removal of the soiling layer is achieved without conduction of heat into the underlying artwork.

As explained in the previous chapter, preferential absorption by the layer of soiling is important for cleaning to proceed selectively. Figure 4.1 shows the absorptivity of polluted and clean surfaces of a Carrara marble sculpture measured at 1.06 μm (Cooper and Larson, 1996). A typical encrustation resulting from atmospheric pollution has an absorptivity of approximately 0.9, i.e. it absorbs 90% of the incident laser radiation, whereas the patina within which the surface detail is preserved has an absorptivity of only 0.3. Once the encrustation has been removed most of the energy in the laser pulse is harmlessly reflected away. Provided cleaning is not carried out at an excessive energy density this difference is sufficiently large for cleaning to become self-limiting. A pure marble surface has an even lower absorptivity of 0.25. The differences in absorptivity between polluted and clean (patina) Carrara marble surfaces at 590 nm (0.9 and 0.6 respectively) and 248 nm (0.9 and 0.8 respectively) are much smaller (Cooper, 1994).

Plate 4.1 shows a conservator working with a laser cleaning system on scaffolding. The laser head, power and cooling supplies are housed in a single mobile unit measuring approximately 60 cm wide by 60 cm tall by 70 cm deep and weighing approximately 125 kg. The system shown in Plate 4.1 has been designed for use in the United Kingdom and is powered from the 13A/240V a.c. mains supply. Laser systems can be configured during manufacture to operate under different conditions. Beam delivery is through a multi-jointed articulated arm which uses mirrors to deflect the laser beam through a series of hollow tubes. The beam emerges through a pen-like handpiece.

An alternative method of beam delivery is the optical fibre which uses the phenomenon of total internal reflection (Wilson and Hawkes, 1983) to guide a beam along a thin flexible glass fibre. Output is again through a ‘pen-like’ handpiece. Both methods of delivery have advantages (Cooper and Loton, 1994). The articulated arm is counterbalanced so that manipulating the beam is very easy and it is also relatively robust. Its reach however is limited. Such a method of delivery is suited to work in a conservation studio or on-site where the object is accessible. The main advantage of the optical fibre is that its reach is limited solely by the length of the fibre. This allows cleaning of less accessible areas since the laser does not have to be close to the object being cleaned. The laser can be left on the ground while a fibre is trailed up to the object. Beam delivery through an optical fibre, however, is not as reliable as delivery through an articulated arm since the high power densities achieved at the entrance and exit faces of the fibre are very close to its damage threshold. This means that occasionally the fibre must be recut and polished.

Within the handpiece a lens is often used to provide the conservator with a diverging laser beam. Occasionally cleaning is carried out with a parallel beam so that the cleaning effect becomes independent of working distance. Working with a parallel beam, however, is potentially more dangerous than a diverging beam since it remains hazardous at much larger distances from the source.

Commercial cleaning systems typically emit several hundred millijoules of energy per pulse in multimode form at repetition rates up to about 30 Hz (a 60 Hz system suited to large-scale cleaning is also available). Some systems also provide the capability of frequency-doubling the fundamental output of the Nd:YAG to give radiation at 532 nm in the green part of the spectrum. Some manufacturers believe that this will increase the versatility of the system. At the present time most cleaning is still carried out at the fundamental wavelength.

Cleaning parameters

The conservator has control over three parameters which affect the cleaning process: the energy of each pulse; the working distance, i.e. the distance between the handpiece and surface being worked on (assuming the beam is diverging); and the repetition rate. The energy per pulse and working distance together determine the *energy density* of the laser beam on the surface. The energy density is a measure of how concentrated the energy is in the beam and is defined by:

$$\text{Energy density } (J/cm^2) = \frac{\text{Energy per pulse } (J)}{\text{Area of laser beam on surface } (cm^2)} \quad (4.1)$$

The energy density determines the effect a laser pulse (fixed wavelength and pulse length) has on a particular surface. If the energy from a single pulse is concentrated into a small area then the beam becomes very intense. If, however, the same amount of energy is spread out over a much larger area then the beam becomes much weaker. Changes in the working distance lead to changes in the beam size and hence to changes in the incident energy density (see Figure 4.2). The following description applies to work beyond the focal point of the lens (work can also be carried out before the focal point which is particularly useful if the laser beam is focused so tightly that it causes ionization of the air at the focal point, visible as a bright spark). When cleaning an object it is desirable to locate the position at which the laser beam just begins to remove the dirt (position 2) and the position beyond which (at shorter distances) the beam begins to damage the object surface (position 1). If the surface is maintained within this 'safe cleaning window' then the cleaning process will be self-limiting. Positions 1 and 2 depend on several factors including the energy of the pulse and the ablation thresholds of the polluted and clean surfaces. The degree of

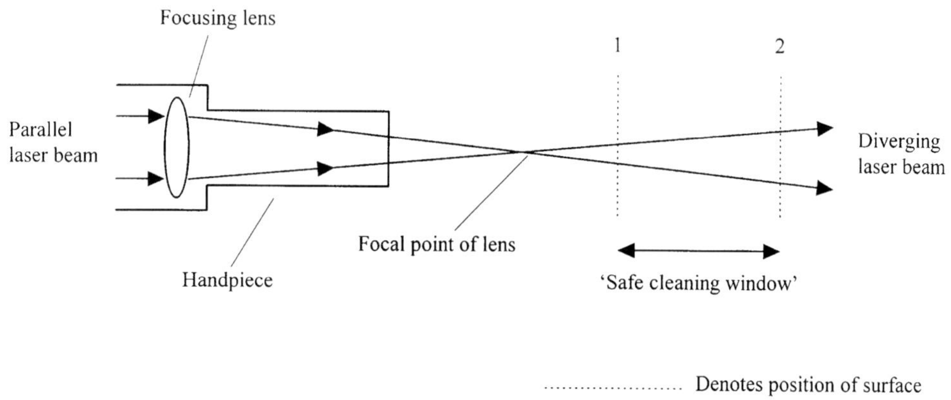

Figure 4.2. *Schematic representation of cleaning with a diverging laser beam.*

cleaning is controlled by the working distance and the number of pulses delivered to the surface. During cleaning it is often necessary to change the beam size without changing the energy density, e.g. work on a fragile intricately carved surface with areas of pigment requires much more precise work than a large flat surface in sound condition. This is achieved by changing the energy per pulse:

Decrease energy per pulse ⇒ work with smaller beam ('safe cleaning window' moved towards the handpiece)

Increase energy per pulse ⇒ work with larger beam ('safe cleaning window' moved away from the handpiece)

If conditions are such that positions 1 and 2 are interchanged, e.g. if the dirt layer is extremely difficult to remove or the artefact is especially vulnerable to laser radiation at 1.06 μm, then the process is not self-limiting and the conservator must avoid exposing the cleaned surface to laser radiation.

The repetition rate at which cleaning is carried out controls the speed of cleaning: the higher the repetition rate, the faster the cleaning. The conservator must bear in mind that at high repetition rates there is less time between pulses and decisions must be made more quickly. If, for example, the conservator is unaware of the existence of pigment beneath a dirt layer and cleaning is carried out at a high repetition rate, e.g. 10 Hz or more, the pigment might be ablated before the conservator realizes it is there. The conservator must strike the right balance between control and speed.

A suitable combination of energy per pulse, working distance and repetition rate allows the conservator to control not only what is removed but also how much is removed from a surface.

Practical cleaning techniques

Laser cleaning should be treated no differently from other techniques available to the conservator and should only be used if it is the most appropriate tool for the job. Cleaning should whenever possible be carried out at the lowest possible energy density so that ablation occurs via the most selective mechanism, thereby minimizing any risk of damage to the artwork. The laser must only be used by trained personnel and all health and safety guidelines must be followed. Also, as with other techniques, records of cleaning procedures should be kept. This is particularly important with laser cleaning being a relatively new technique. It must be remembered that careless and inappropriate use of the laser, just like any tool, will result in damage to an artwork.

Test cleaning

As with other techniques, cleaning tests should be carried out to establish the most suitable cleaning parameters before proper cleaning is started. Such tests should be carried out if possible on an inconspicuous area of surface which is representative of the surface as a whole, the aim being to establish the minimum energy density, or range of energy density (cleaning is often undertaken at several levels), which gives the desired level of cleaning without damaging the artwork. The beam size and repetition rate will be determined by the condition and intricate nature of the surface, by the presence of other materials on the surface – remnants of pigment or old repairs – and by the experience of the conservator. Since the level and type of soiling and condition of the artwork often change from one area of the surface to another it is usually necessary to adjust the cleaning parameters during cleaning. The conservator must ensure that an area cleaned at a different energy density blends into the rest of the surface.

Indications of damage during cleaning

On some surfaces damage caused by cleaning at excessive energy density is visible as changes to the topography, such as surface pitting, or changes to the colour, e.g. absorption of energy by certain pigments leads to oxidation. On other surfaces, particularly irregular surfaces, damage is much harder to detect visually. However, damage to any surface is usually accompanied by the generation of light (in the form of a plasma) and sound during the laser beam/surface interaction.

The removal of dirt from a surface is accompanied by an audible 'snapping' sound and the emission of light by ejected material just above the surface. Once the dirt layer has been removed, provided the cleaning process is self-limiting, the sound and light disappear since material is no longer ejected. In some cases a low amplitude acoustic pulse and low intensity flash continue to accompany each laser pulse as a result of absorption by impurities in the surface. An indication that the process is not self-limiting is the continued emission of sound and light with each laser pulse after the dirt has been removed. Figure 4.3 shows a graphical representation of the loudness of the acoustic pulse generated during ablation as a function of number of laser pulses for polluted and clean limestone surfaces (Cooper, 1994). The decrease in amplitude from the polluted surface corresponds to cleaning of the stone surface. After approximately ten pulses the crust has been completely removed and the acoustic pulse amplitude is only 1% of its initial value.

On some surfaces, such as marble, laser damage is indicated by the generation of a bright spark on the surface. Such damage usually takes the form of surface pitting. The formation of an intense plasma just above a polluted surface during cleaning indicates that ablation is proceeding via less selective mechanisms which can result in damage to the underlying surface.

The generation of sound and light during the cleaning process can be used by the conservator as a guide to its selectivity. Feedback such as this is helpful in avoiding damage to the surface being worked on.

Steam laser cleaning

The use of water with laser cleaning is known as *steam* laser cleaning. As explained in the previous chapter, the application of a thin film of water to a surface immediately prior to irradiation, either by fine spray or brush, often enhances the cleaning effect (see Figure 3.10). This is especially useful when (a) the soiling is bonded very strongly to the surface and cannot be removed by dry laser cleaning, and (b) the surface of the artefact is fragile and cleaning must be carried out at reduced energy density. Plate 4.2 demonstrates the enhanced cleaning effect of steam laser cleaning. Steam laser cleaning also usually increases the rate of cleaning.

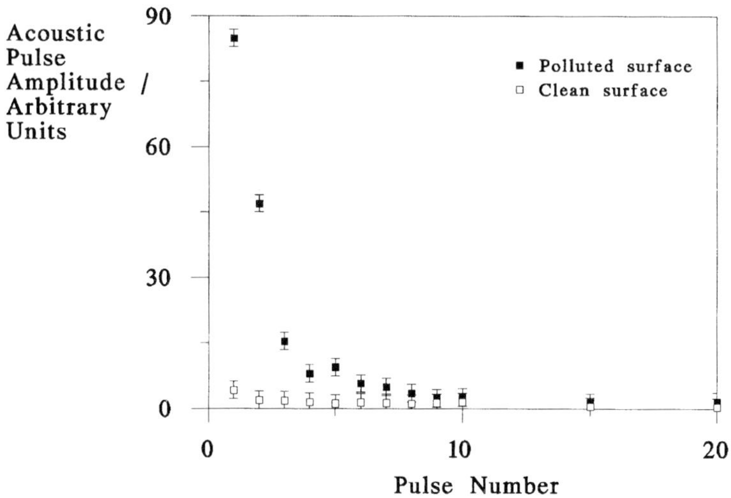

Figure 4.3. *Acoustic pulse amplitude as a function of number of pulses for clean and polluted limestone surfaces (Cooper, 1994).*

Tests should always be carried out beforehand to establish the suitability of using water. On some surfaces the use of water can lead to staining or the formation of tide marks, and an area which has been steam laser-cleaned can sometimes be difficult to blend with an area which has been dry laser-cleaned. Unless great care is taken a surface cleaned by different laser cleaning techniques adopts a patchy appearance. As a general rule water should only be used if needed.

Keeping records

Record keeping is important so that future generations of conservators can find out exactly how an object has been treated in the past. Knowledge of past treatments influences the way in which conservation work is undertaken. Also, the more a conservator knows about the likely effects of laser radiation at a certain energy density on a surface, the more likely they are to use laser cleaning in an appropriate manner. Just as a conservator using chemicals to clean a surface keeps a record of the type of chemical, its concentration and method of application, a conservator using a laser should record the wavelength, pulse length and energy density used (repetition rate, energy per pulse and beam size can also be useful).

In order to calculate the energy density both the energy per pulse and beam size are needed. Most systems have an in-built energy monitor from which the conservator can note the energy per pulse. Beam size can be measured in a number of ways (no standard method of measurement has been adopted) but the most practical in a conservation environment is to use photosensitive or 'burn' paper. The beam shape is recorded by placing a small piece of burn paper in the way of the laser beam at the surface of the object. Strong absorption of the energy in the laser pulse ablates the surface of the paper, leaving an imprint of the beam. From this its size can be estimated. Division of energy per pulse by beam size gives a value for the *average* energy density. It is important to note that in reality a laser beam is not perfectly uniform, i.e. the energy is not distributed perfectly evenly across the beam, and that certain parts of the beam known as *hot spots* are more intense than others. It is at these hot spots that damage is most likely to occur. The method outlined here only gives an approximate average value of the energy density, believed to be accurate to within ±25%, but it does have the virtues of being simple and practical. Provided the conservator adopts a consistent approach this accuracy is adequate for record keeping. With some laser cleaning systems the conservator uses a spacer attached to the handpiece to maintain a fixed working distance so that the energy density on the surface matches some pre-determined value.

Advantages and disadvantages of laser cleaning at 1.06 μm

Laser cleaning at 1.06 μm has a number of advantages over more conventional cleaning methods:

- *Selectivity:* Radiation at 1.06 μm is preferentially absorbed by many types of soiling and surface accretions. This means that, within a range of parameters, laser cleaning is self-limiting and soiling can be removed from the surface of an artwork in an extremely selective manner; tool markings are often revealed and delicate patinas left intact. It is also possible to go back over areas which have already been cleaned and 'pick off' remnants of dirt without overcleaning the surrounding areas.
- *Precision and versatility:* The laser cleans only where the beam is directed and the beam size is variable between one millimetre and several centimetres. The same tool can, therefore, be used to clean small scale intricate carving and lifesized sculpture. The cleaning effect is also variable: at high energy density the most stubborn dirt can be removed while at low energy density the beam will 'wash' gently over a surface. The same laser can also be used to remove a wide range of surface accretions from a wide range of materials including different types of stone, terracotta, aluminium, bone, ivory, wood and parchment.

- *Non-contact:* The use of light as a cleaning medium, which makes no mechanical contact with the surface, allows badly-decayed crumbling surfaces to be worked on.
- *Immediate control and feedback:* The cleaning action stops as soon as the laser is turned off. Visual information about the state of the surface is continuously fed back to the conservator since there is no cleaning medium to obscure the surface. This means that decisions concerning the treatment can be made at an earlier stage.
- *Reliability:* The Nd:YAG laser is an extremely reliable tool. There are no moving parts within the laser head so components are not subject to general wear and tear. Simple maintenance at regular intervals is sufficient to keep the laser operating efficiently.
- *Environmental:* Laser cleaning is a 'clean' process. There is no use of hazardous chemicals or solvents and the only waste product generated is the dirt ejected from the surface which is relatively easy to collect with an efficient extraction system. The only protective clothing necessary for the conservator is a pair of safety spectacles and a face mask. Set-up times are therefore kept to a minimum.

Laser cleaning at 1.06 μm does have a number of disadvantages:

- *Expense:* The high initial cost of a laser cleaning system is seen by many as a major disadvantage. However, this should be balanced against the reliability of the system (low running costs) and the ability it gives the conservator to complete many jobs in a shorter time, carry out projects which are not possible using more conventional techniques and, most importantly carry out work to a higher standard. It is possible to clean a lifesized medieval alabaster effigy, for example, in 2–3 days using a laser rather than in the 2–3 weeks it would take a conservator using solvents. With such time savings a laser system could pay for itself within a few years. In the short term lasers can be hired for specific jobs or leased for longer periods.
- *Ineffective on some surfaces:* Although the Nd:YAG laser is a versatile tool laser radiation at 1.06 μm is unsuitable for some surfaces, e.g. certain types of algae and lichen are burned rather than ablated, some pigments are sensitive to radiation delivered even at low energy density and absorption of energy by old repairs can lead to their discoloration or weakening. Some minerals, such as mica and red iron oxide, absorb strongly at 1.06 μm. The cleaning of granite and red Glasgow sandstone must, therefore, be carried out very carefully to ensure that minerals such as these are not altered or removed upon irradiation. Different wavelengths are being looked at as possible solutions to these problems. Laser systems are already available which can provide laser radiation in both the near infrared and visible parts of the spectrum. Within a few years this range will have been extended into the ultraviolet. By how much systems such as these will improve the versatility of the laser as a conservation tool is not yet clear.

For many conservators the advantages of laser cleaning clearly outweigh the disadvantages and introduction of the technique has been seen as a significant step in the refinement of conservation procedures. Quite simply, the level of precision and control offered by the laser has never previously been available to the conservator.

Health and safety

The aim of this section is to provide the interested reader with a general understanding of the basic safety precautions which must be followed when using a laser. It does not attempt to provide the conservator with a rigid health and safety framework for their workplace. Situations in which lasers are used often differ and health and safety guidelines should be drawn up by the institution where the laser is being used, in collaboration with the laser supplier and in accordance with current legislation. Anyone coming into contact with a laser must be aware of its potential hazards and must adhere to the health and safety guidelines put in place to minimize these. Training is not only vital in terms of conservation practice, but most importantly in terms of accident avoidance.

The characteristics of laser radiation give rise to two major safety problems (Carr, 1972):

1. the laser beam can traverse large distances with little change in its characteristics, and
2. the laser is a source of extreme brightness (several orders of magnitude brighter than the Sun).

The eye is a sensitive light detector and is, therefore, especially vulnerable to laser radiation damage, either at the retina due to focusing by the lens or at the cornea due to direct absorption. In both cases the eye can be painfully and irreparably damaged. At infrared wavelengths laser radiation is transmitted by the eye and damage is caused by the focused laser beam on the retina (see Figure 4.4). In the deep ultraviolet damage is caused to the cornea. Infrared and ultraviolet radiation are also absorbed by the skin: infrared radiation is converted to heat and causes burning and ultraviolet radiation can induce photochemical reactions, the general effect of which is that of severe sunburn.

Great care must be taken to ensure the safety of the conservator using the laser and of people nearby. Laser cleaning systems used by conservators are class IV products, that is potentially the most hazardous type of laser system, and must be treated with the utmost respect. A laser safety programme should be drawn up at institutions where a laser is being used (manufacturers of systems should be able to provide assistance) and safety guidelines must be strictly followed. The laser safety programme is divided into three parts: administration controls, engineering controls; and personal protection.

Administration controls are set up to ensure that only authorized personnel use the laser. It is essential that all people using a laser are aware of the potential hazards and of the subsequent control measures needed. Before using the laser conservators must attend a training programme, the goal of which is to give them an appropriate and realistic understanding of the tool with which they will be working. The conservator does not have to understand laser physics but an understanding that a laser can cause serious eye damage even at large distances is vital. *Engineering controls* are necessary to ensure that laser cleaning is carried out in an enclosed environment with restricted access. Points of access must be interlocked so that the laser is switched off automatically as soon as anyone enters the area and must have a warning light which is activated when the laser is on. No other work should be carried out in the same area while the laser is in operation. In a conservation studio laser safety guidelines are straightforward to follow if cleaning is carried out in a designated room. If cleaning has to be carried out in an open area, e.g. a large studio or outside, the same procedures must be followed and temporary screening must be erected around the object to form an enclosed working

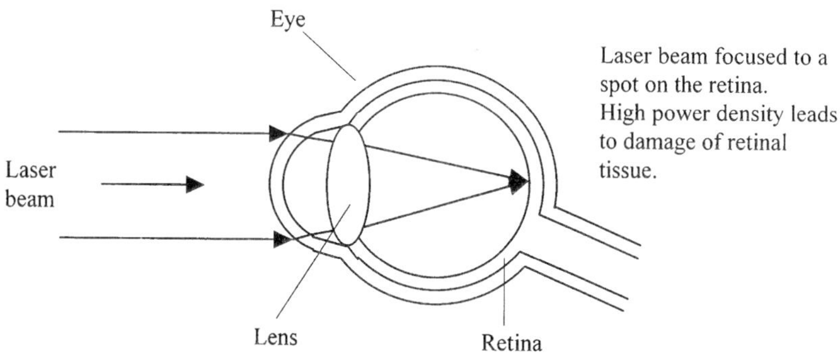

Figure 4.4. *Focusing action of the eye at infrared wavelengths. Transmission of laser radiation leads to damage of the retina.*

area. *Personal protection* in the form of eye protection, face masks and extraction systems forms an important last line of defence. Eye protection must always be worn when working with a laser and also by anyone watching. Such protection is available as glasses or goggles, which must be suitable for the wavelength being used and must fit correctly. Efficient extraction is essential to remove material ejected from the surface during cleaning (necessary from a health point of view and to prevent it settling back onto the surface). Since extraction systems are not one hundred per cent effective a face mask should also be worn by the conservator to prevent inhalation of the particulates and any vapours produced during the process. The products generated during laser cleaning have not yet been extensively studied but initial research into the products of cleaning mould-contaminated paper (Friberg, Zafiropulos, Kalaitzaki *et al.*, 1997) and research carried out in other laser processing areas have suggested that the particulates and vapours generated during the process can pose a serious health hazard to the operator in the long term (Watkins, in press).

Neglect of any one of these areas creates a serious weakness in the safety programme. Since the long-term health implications of laser cleaning are unknown additional guidelines are often adopted to protect conservators. At the National Museums and Galleries on Merseyside (NMGM) the guidelines state that a conservator should take a break of at least five minutes after working with a laser for 30 minutes to help prevent eye strain caused by intense concentration on the surface. These additional guidelines should be flexible since people have different tolerance levels and can sustain concentration for different lengths of

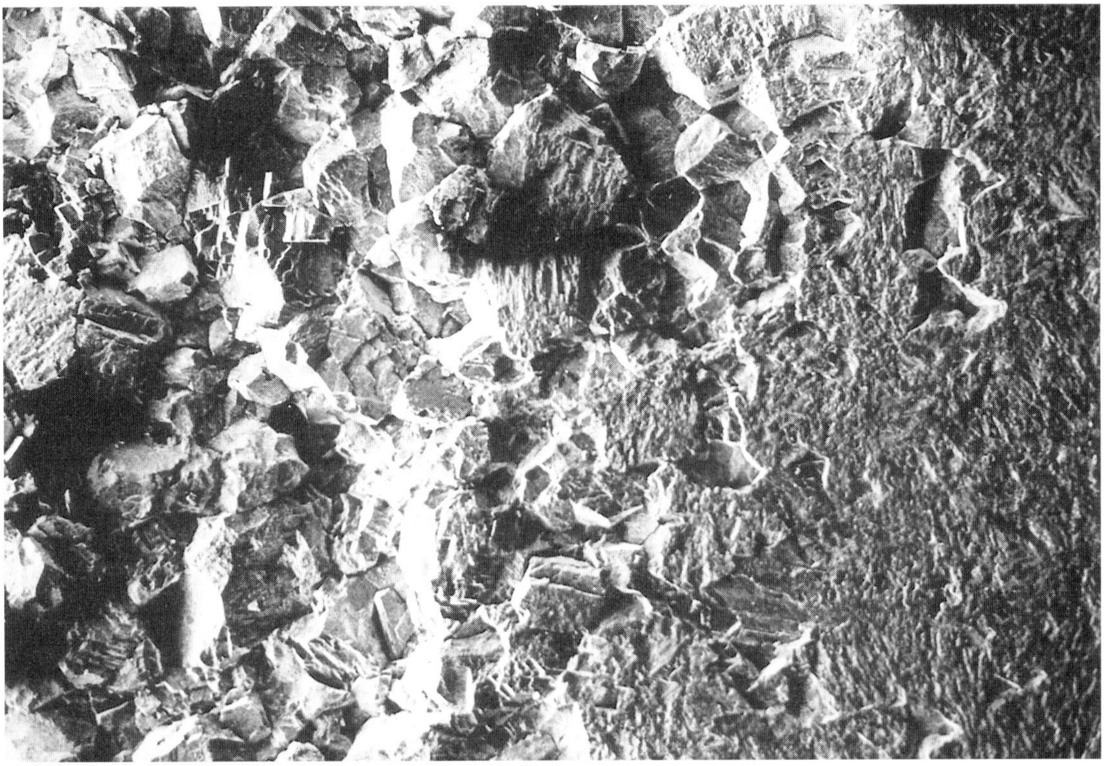

Figure 4.5. *Electron micrograph of a cleaned marble surface (4 × 3 mm). The left half has been cleaned by Q-switched laser radiation at 1.06 μm at an excessive energy density. Crystals have been 'knocked out' of the surface leaving craters approximately 300 μm wide. The right half shows the clean, undamaged marble surface for comparison.*

time. This section is certainly not meant to frighten conservators away from the laser but is simply meant to draw attention to the potential hazards of using laser radiation and to the procedures which are necessary to prevent accidents. Provided an effective laser safety programme is in place and the conservator understands the potential dangers of laser radiation accidents are very unlikely to occur.

Summary

Laser radiation at 1.06 μm provides the conservator with an extremely precise and selective method of removing a wide range of surface accretions from a wide range of materials (controlled removal of unwanted layers is also provided by ultraviolet laser radiation and is dealt with in Chapter 6). It must be stressed that a laser used carelessly or inappropriately can, just like any tool, damage a wide range of materials also (see Figure 4.5). The laser simply provides the conservator with another tool. Sometimes it is necessary to use a range of cleaning techniques on an object. The laser should not be used to the exclusion of all other methods unless it is the most appropriate tool for all aspects of the job. In order to realize the full potential of the laser conservators should be trained in both practical and basic theoretical aspects of laser cleaning. This is also important from a safety point of view. Laser training courses are available and typically last a few days. The quality of cleaning achieved using a laser depends on the judgement of the conservator and the more they understand about the processes involved the more likely cleaning will be of high quality.

References

Carr, J.F. (1972) Laser safety. In *Lasers in Industry* (S.S. Charschan, ed.), pp. 579–609, Van Nostrand Reinhold Company

Cooper, M.I. (1994) *Laser Cleaning of Stone Sculpture*. PhD Thesis, Loughborough University

Cooper, M.I. and Larson, J.H. (1996) The use of laser cleaning to preserve patina on marble sculpture. *The Conservator*, **20**, 28–35

Cooper, M.I. and Loton, A. (1994) Light years ahead? *Natural Stone Specialist*, December, 24–30

Friberg, T.R., Zafiropulos, V., Kalaitzaki, M, *et al.* (1997) Excimer laser cleaning of mold-contaminated paper: sterilization and air quality considerations. *Lasers in Medical Science*, **12**, 55–59

Watkins, K.G. (1997) A review of materials interactions during laser cleaning in art restoration. *Restauratorenblatter*, Proceedings of LACONA I (Crete, October 1995), International Institute for Conservation of Historic and Artistic Works, Verlag Mayer, Vienna

Wilson, J. and Hawkes, J.F.B. (1983) *Optoelectronics: An Introduction*, Prentice-Hall International, pp. 332–76

5

Case studies

Martin Cooper

This chapter describes some of the work which has been carried out during the past three years by the National Museums and Galleries on Merseyside (NMGM) where lasers are now used on a routine basis. Case studies are used to illustrate the kind of problems for which laser cleaning can be the most effective and appropriate technique available to the conservator. Unfortunately, since every laser has slightly different characteristics and every object poses different problems to the conservator, it is not possible to produce a booklet instructing the conservator which values of energy density, repetition rate etc. to use for different materials. Every object must be treated independently. Laser cleaning should only be used if it is the most appropriate technique and, as with other techniques, tests should always be carried out initially to establish the parameters required for the desired level of cleaning. Hopefully, by the end of this chapter the reader will have a better understanding of when and how a laser can be used to overcome certain conservation problems. It is important to remember that the quality of cleaning achieved using the laser is determined by the skill and experience of the conservator.

The work at NMGM has been carried out by conservators using a multimode Q-switched Nd:YAG laser operating at a wavelength of 1.06 μm with a pulse length of 5–10 ns, a maximum pulse energy of 250 mJ, a maximum repetition rate of 10 Hz and a beam diameter of typically 5–10 mm (in cases where work had to be carried out very precisely a much smaller beam was used). The average energy densities quoted have been calculated by dividing the pulse energy by the beam size (as measured with burn paper) and are accurate to within ±25%. These are meant as general guidelines only since, in reality, cleaning is usually carried out at a range of values. Unless stated otherwise the energy density quoted is the value at which the majority of cleaning was undertaken. At NMGM most experience has been gained from cleaning marble, limestone and terracotta sculpture. More recently, as the advantages of laser cleaning have become clear, the areas of work have broadened to include a much wider range of materials: bone, ivory, parchment, plaster and aluminium. In some of these areas work is still at an initial stage but in many Q-switched Nd:YAG laser cleaning has become a valuable technique for the conservator. Case studies involving ultraviolet laser cleaning are presented in Chapter 6.

Marble sculpture

The cleaning of a wide range of marble sculpture by laser radiation has met with considerable success. Laser cleaning has given the conservator the ability to clean marble sculpture without damaging its patina (Cooper and

Larson, 1996; 1997). In this section five case studies are presented: four illustrate the effectiveness of laser radiation in removing dirt layers whilst preserving different forms of patina, and one illustrates a situation where use of the laser was found to be inappropriate.

A nineteenth-century statue of William Huskisson

Plate 5.1 shows a nineteenth-century over-lifesized Carrara marble statue of a former Liverpool politician, William Huskisson. The statue was in poor condition as a result of over 100 years' exposure to a polluted atmosphere and was brought indoors for cleaning and repair. A hard black pollution crust, 0.1–1.0 mm thick, covered much of the surface and in damaged areas the fine-grained marble had lost its cohesion and adopted a sugary texture. Cross-sectioning of a sample taken from a polluted area of the surface revealed three distinct regions (see Plate 5.2): a hard black dirt layer about 0.5 mm thick, a pale yellowish-brown patina about 0.2 mm thick and the underlying calcite crystals of the marble. Within the patina, formed by reaction of the marble surface with acidic gases in the polluted atmosphere, is the original surface relief of the statue. The yellowish-brown coloration is staining resulting from incorporation of organic materials during formation of the patina. The surfaces of the calcite crystals have been etched by the weathering process.

The statue of William Huskisson has been used to compare the effect of laser cleaning with more conventional methods of cleaning. Partial cleaning of the statue by steam and poulticing methods has left the surface with a raw and patchy appearance (see Plate 5.3, right area); in some areas the patina has been removed along with the dirt revealing the etched calcite crystals, while in other areas stubborn deposits of dirt remain. Partial cleaning of the statue with laser radiation, carried out with water at an average energy density of 0.4 J/cm^2, has revealed a smooth and undamaged patina from which the dirt layers have been completely removed (see Plate 5.3, left area). Tool markings have been preserved in the surface of the patina. A thin-section taken through a partially-cleaned sample from the statue (see Plate 5.4) clearly shows that the dirt layers have been removed by the laser without disrupting the patina.

Figure 5.1 shows a comparison between an area cleaned by laser radiation and a similar area cleaned by air-abrasive. The boundary between the black crust and the patina revealed by laser cleaning is very distinct (Figure 5.1a). The boundary between the crust and the abrasive-cleaned area is less well defined (Figure 5.1b) and there is strong evidence of overcleaning: a qualitative elemental study of the cleaned surface revealed it to be composed mainly of calcite, rather than calcium sulphate, crystals, i.e. the patina had been removed. The surface is also very uneven, due to the impact of abrasive particles. Removal of pollution encrustations by abrasive cleaning often leaves the stone surface with a mottled appearance, composed of peaks and troughs with black specks of dirt left sitting on top of the peaks. Since a stream of abrasive particles does not discriminate between dirt and stone, some areas of the surface are overcleaned while others are undercleaned (see Figures 1.1, 1.2). The selectivity of laser radiation at 1.06 μm allows relatively thick dirt layers and pollution encrustations to be removed from marble surfaces without overcleaning. The conservator is able to return to a lightly cleaned surface and remove any remnants of dirt without damaging the surrounding surface.

In this case the laser provided the conservator with a means of both cleaning a heavily polluted marble sculpture and of preserving its valuable patina. The statue of William Huskisson is now an exhibit in the NMGM Conservation Centre, illustrating the effects of different cleaning techniques.

An Ancient Greek sculpture of a seated girl

Plate 5.5 shows an Ancient Greek marble sculpture of a seated girl (owned by the British Museum). The sculpture was carved from a large-grained Greek marble, the surface of which has decayed badly over 2000 years leaving a very granular, porous surface. The dirt particles were consequently bound very

Plate 1.1. *Cleaning tests on a nineteenth-century architectural terracotta panel from the Mercantile Mariner's Association in Wallasey, Merseyside. The left side has been cleaned by Q-switched Nd:YAG laser radiation at a wavelength of 1.06 μm. Two patches on the right side have been steam-cleaned. The areas cleaned by steam have a much deeper red coloration, which has been attributed to an enrichment of the iron concentration at the surface due to water penetration into the terracotta (Larson and Cooper, 1996).*

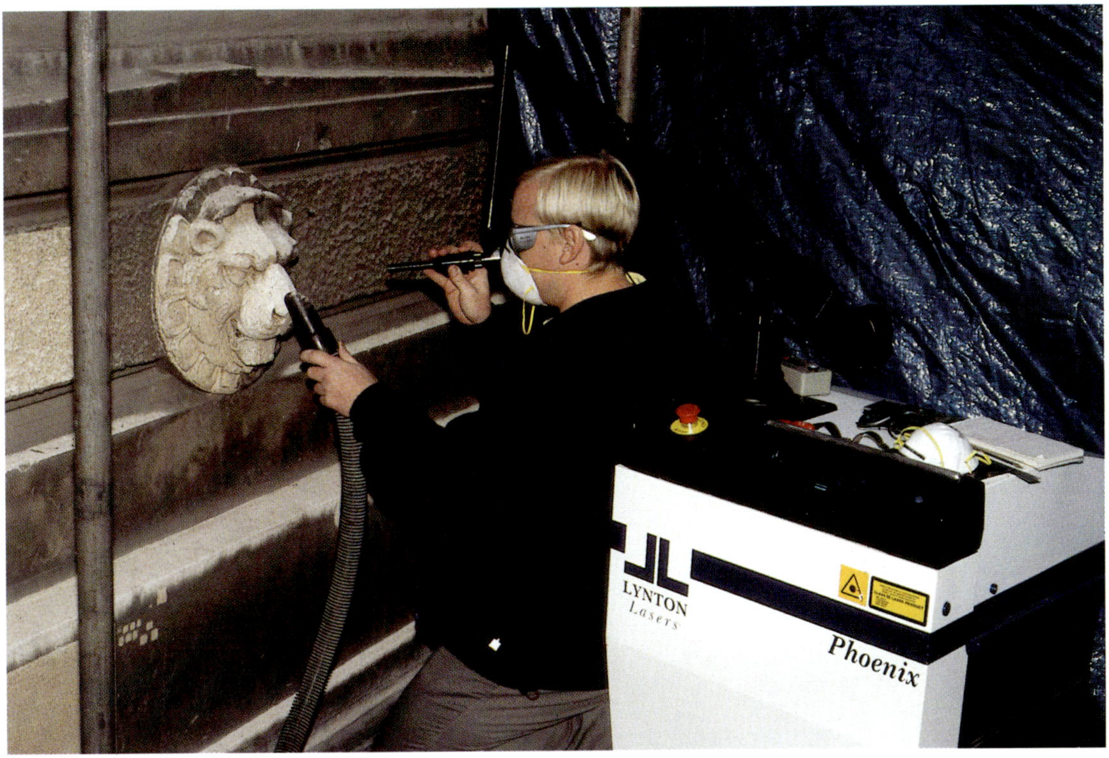

Plate 4.1. *Laser cleaning of sculptural detail on the outside of a building.*

Plate 4.2. *The enhanced effect of steam laser cleaning. The lower part of the terracotta decoration has been cleaned by Q-switched Nd:YAG laser radiation at approximately 0.3 J/cm²; (a) left half with water, (b) right half without water.*

Plate 5.1. *A polluted nineteenth-century marble statue of William Huskisson.*

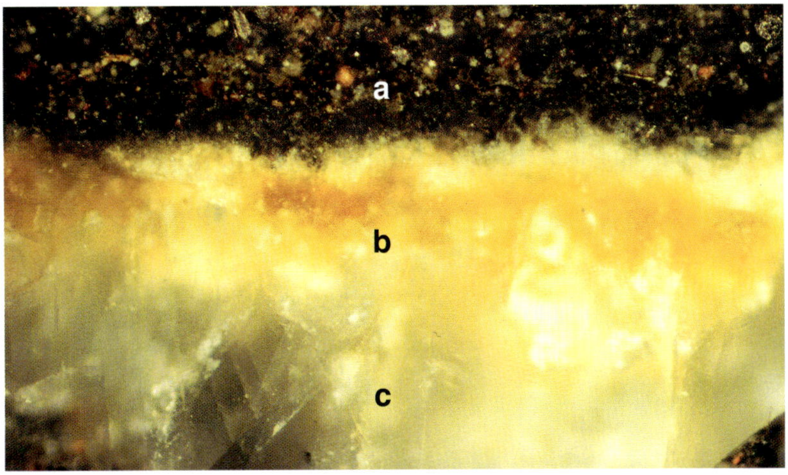

Plate 5.2. *Cross-section through a polluted sample from the statue of William Huskisson; (a) black crust (0.5 mm thick), (b) yellowish-brown patina (0.2 mm thick) and (c) calcite crystals of the marble.*

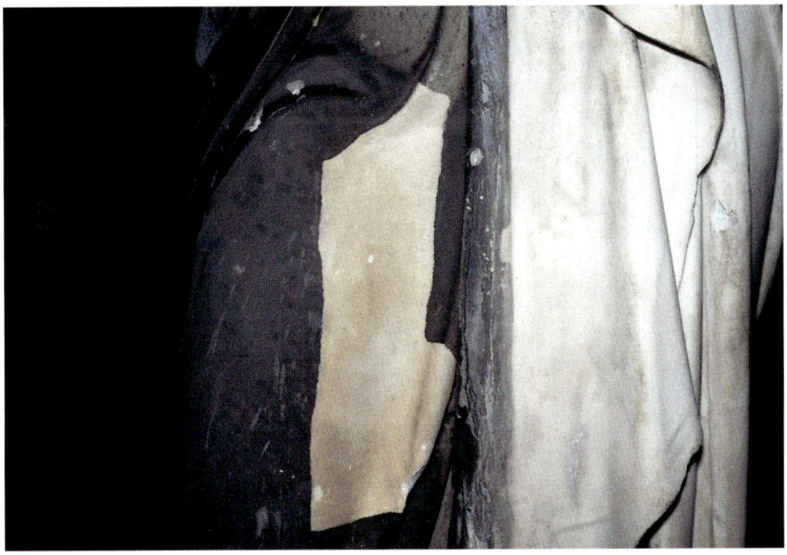

Plate 5.3. *A detail from the drapery of the William Huskisson statue showing a steam and poultice-cleaned area (right side) and a laser-cleaned area (left side)*

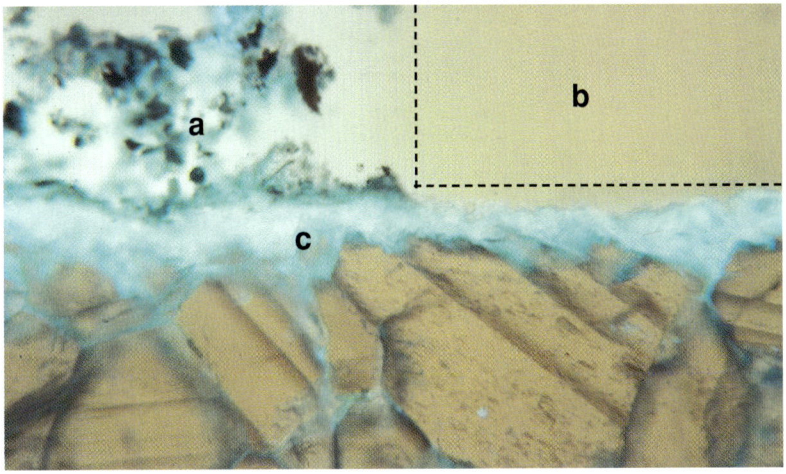

Plate 5.4. *A thin-section through a partially cleaned sample from the William Huskisson statue, viewed under ultraviolet light. Left side (a) shows the porous dirt layer, right side (b) shows the laser-cleaned surface. The light blue layer (c) is the patina (0.2 mm thick).*

Plate 5.5. *An ancient Greek marble statue of a seated girl (50 cm tall, owned by the British Museum).*

Plate 5.7. *The statue of a seated girl after completion of laser cleaning.*

Plate 5.6. *The statue of a seated girl after initial laser cleaning (left side)*

Plate 5.8. *A Hellenistic marble head (28 cm tall). The right side has undergone initial laser cleaning.*

Plate 5.10. *The Hellenistic head after completion of laser cleaning.*

Plate 5.9. *The Hellenistic head after completion of initial cleaning. The white patch around the left eye has undergone a final stage of cleaning to remove the limewash and remaining soil particles.*

Plate 5.11. *A seventeenth-century marble bust of Goliath. The left side has been laser-cleaned, revealing a uniform yellow-brown patina.*

Plate 5.12. *A classical marble sculpture with eighteenth-century restorations before cleaning.*

Plate 5.13. *Laser test cleaning areas on the back of the classical marble sculpture. The central area (1 cm²) was cleaned at low energy density (b), the left area at high energy density (a).*

Plate 5.14. *A heavily polluted grotesque (limestone, 70 cm tall) from Lincoln Cathedral. Reproduced by kind permission of the Dean and Chapter of Lincoln.*

Plate 5.16. *A fourth-fifth century AD Egyptian limestone relief (55 cm × 45 cm). Initial cleaning has been carried out on the left side.*

Plate 5.15. *Area on the lower front of the grotesque which has been laser-cleaned.*

Plate 5.17. *An area of the Egyptian relief after completion of cleaning and consolidation.*

Plate 5.18. *Architectural detail (sandstone, 30 cm wide) half cleaned by laser.*

Plate 5.19. *Blonde sandstone ashlar cleaned by laser (cleaned area is 30 cm wide).*

Plate 5.20. *A seventeenth-century over-lifesized terracotta statue of Neptune. The right side has been laser-cleaned.*

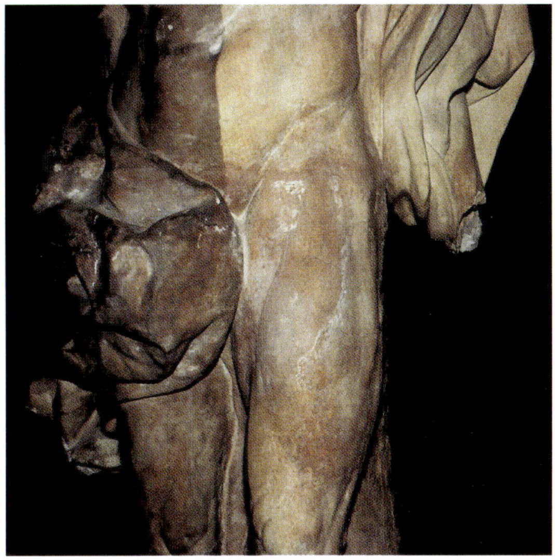

Plate 5.22. *A close-up of an area of original terracotta wash (right side) on the statue of Neptune revealed by laser cleaning.*

Plate 5.21. *A close-up of the partially cleaned statue of Neptune.*

Plate 5.23. *Architectural detail (terracotta, 30 cm wide) half cleaned by laser.*

Plate 5.24. *Plaster model (1.25 m tall) for the Port Sunlight war memorial. The left side has been laser-cleaned.*

Plate 5.25. *A close-up of the plaster model showing areas of dark shellac cleaned by laser radiation (left side).*

Plate 5.26. *Section of drapery from the Eros statue partially cleaned (right side).*

Plate 5.27. *A late nineteenth-century ivory and bronze sculpture, Mors Janua Vitae, by Harry Bates.*

Plate 5.28. *Small ivory figures (approximately 11 cm tall) on the bronze base of the Bates sculpture: (a) before cleaning and (b) after laser cleaning.*

(a)

(b)

Plate 5.29. *Egyptian archaeological ivory artefact (approximately 7 cm × 5 cm) after laser cleaning. Original soiling has been left for comparison.*

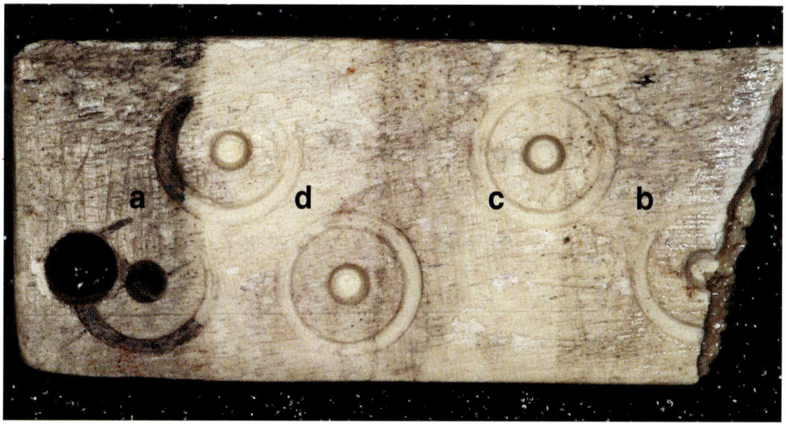

Plate 5.30. *Egyptian archaeological bone artefact (approximately 3 cm long) showing different levels of cleaning: (a) original soiled surface, (b) 0.2 J/cm², (c) 0.5 J/cm² and (d) 0.7 J/cm².*

Plate 5.31. *Script on nineteenth-century parchment (area is 3.5 cm × 2.5 cm). The left side has been laser-cleaned without removal of the script.*

Plate 5.32. *A painted oak misericord (30 cm wide).*

Plate 5.33. *Attempted layer-by-layer paint removal from the misericord by laser (treated area is 2.5 cm × 2.5 cm). The uncontrolled nature of the process has led to an extremely uneven result; different paint layers have been removed in different places.*

Plate 5.34. *The back of the misericord (area is approximately 14 cm × 10 cm). A thin layer of dirt has been removed from the oak surface by laser radiation on the right side.*

Plate 6.1. *Uncovering of a seventeenth-century painting from beneath a nineteenth-century overpainting using a KrF excimer laser. Left: one step before the uncovering. Right: after completion of the treatment.*

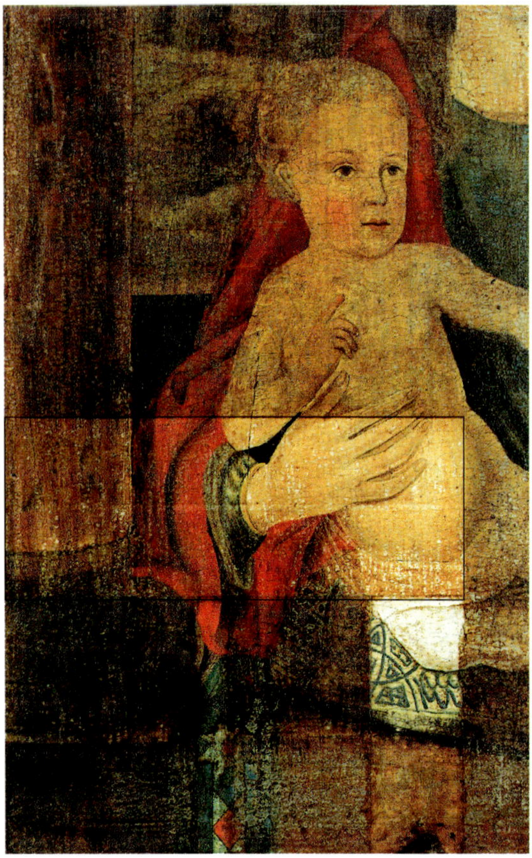

Plate 6.2. *Varnish removal using KrF excimer laser (middle section) and chemical solvents (two zones at bottom).*

Plate 6.3. *Segment of overpainted plaster. Removal of overpaint using a KrF excimer laser has been performed on two areas; the original colours have been uncovered.*

Plate 6.4. *Trials of aged varnish removal aided by on-line control based on LIBS spectra of Figure 6.7.*

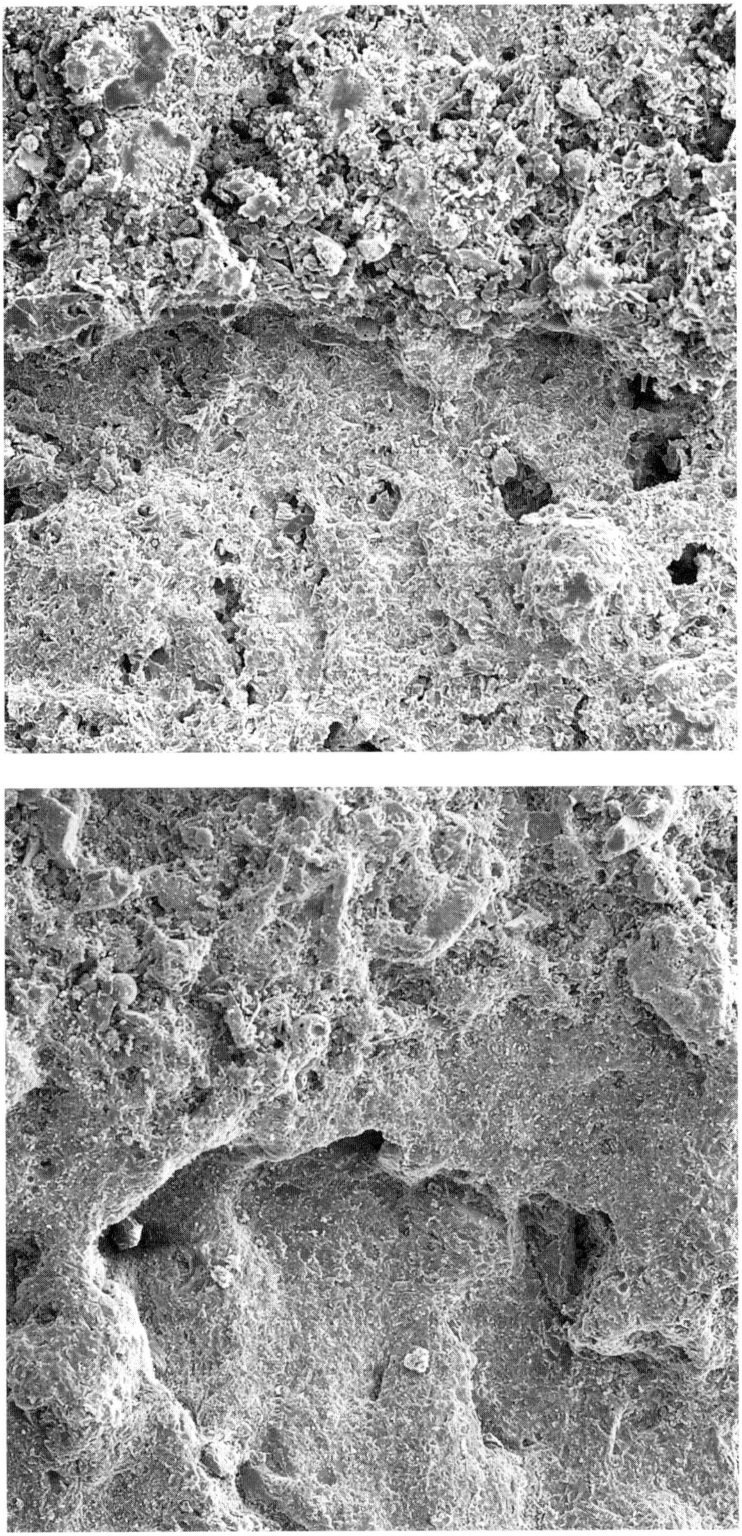

Figure 5.1. *Electron micrographs of partially (a) laser-cleaned and (b) abrasive-cleaned surfaces of the William Huskisson statue; black crust (upper half), cleaned surface (lower half). Scale: area is 1 mm × 1 mm.*

(a)

(b)

strongly within the surface, between the marble crystals, and cleaning by conventional methods either proved unsuccessful in removing the dirt or was not attempted for fear of damaging the marble.

An initial clean, carried out at an energy density of 0.3 J/cm^2, was successful in removing most of the dirt without damaging the patina (Plate 5.6, left side). The stubborn deposits of dirt were then carefully removed using water and a slightly higher energy density of 0.4 J/cm^2. Lichen which was attached strongly to some damaged areas of the granular surface was only partially removed by this cleaning. These dark patches, although small, were very distracting. It was therefore decided that they should be removed and that the laser was the most precise and, hence, most suitable tool for the job. Removal of the organic remnants was successfully carried out with the conservator working under magnification at a relatively high energy density of 2.0 J/cm^2 (the process was not self-limiting). The lichen was 'picked off' using a small beam size of 1.5 mm diameter and a low repetition rate of 2–3 Hz. Plate 5.7 shows the sculpture after cleaning had been completed. In this case the laser provided the *only* safe means of cleaning the sculpture.

A Hellenistic head

Plate 5.8 shows a Hellenistic head of a woman dating to the second century BC. The head is of Parian marble, originates from the eastern Greek islands or western Turkey and is a remnant of an over-lifesized statue. The head was discovered during archaeological clearance work in advance of restoration work at Badger Dingle, Shropshire, England. The marble surface was covered by soil and in some places by remnants of limewash (Plate 5.8, left side). Close examination of the surface revealed that the limewash covered some of the damaged areas indicating that it was not an original application. Cleaning of the head was essential since much of the carved detail was obscured by earth.

Removal of the hard dirt layer by scalpel was ruled out on the grounds that it would be very risky for the marble surface and extremely time consuming. Water and chemical treatments were also ruled out because these would damage any traces of pigment that might remain and might stain the marble. Initial cleaning (Plate 5.8, right side) was carried out in two parts: firstly, at an energy density of 0.4 J/cm^2 with water to remove most of the soil (only a very thin film of water, which was vaporized very quickly, was used so staining was not a problem) and, secondly, at 0.75 J/cm^2 with water to remove the more strongly bound areas of soil. This initial stage of cleaning removed most of the soil but left the limewash untouched. The surface retained a light brown coloration due to soil particles trapped between the large crystal grains of the marble. A final stage of cleaning was carried out at a higher energy density, 1.5 J/cm^2, to 'pick-off' the remnants of limewash and soil (Plate 5.9). Since the process was no longer self-limiting at this level cleaning was carried out under magnification, using a small beam size (2 mm diameter) and low repetition rate (1–2 Hz). Cleaning of the sculpture has revealed the crisp detail while retaining the soft yellow tone of the patina (Plate 5.10).

The laser provided the conservator with a means of carefully working through the soil to uncover the marble surface beneath, the level of cleaning being determined by the conservator.

Head of Goliath

Plate 5.11 shows a seventeenth-century marble bust of Goliath which formed part of a wall fountain. Its surface has a uniform yellow-brown coloration which is possibly of microbiological origin and a result of its prolonged exposure to a damp environment. The surface of the sculpture was covered by a thin (approximately 0.1 mm) grey layer of particulate dirt. Laser cleaning was chosen since the other techniques available (steam, air-abrasive and chemicals) were unable to preserve the yellow-brown patina in its uniform state. Cleaning was undertaken at an energy density of 0.5 J/cm^2. The dirt layer was completely removed revealing a uniform patina undamaged by the cleaning process (Plate 5.11, left side).

A classical sculpture with eighteenth-century restorations

Plate 5.12 shows a classical marble sculpture which was restored heavily during the eighteenth century. As part of the restoration process employed in the seventeenth and eighteenth centuries the surfaces of sculpture were stained using substances such as tea, coffee and tobacco to artificially age any 'new' marble repairs and give the sculpture a uniform antique appearance. Aggressive cleaning since then has removed much of the patina and left the surface with an unattractive patchy appearance. The marble surface was in good condition but covered by a thin grey layer of greasy dirt, a result of exposure to a smoky indoor atmosphere. The aim of cleaning was to remove the dirt layers and artificial patina (where remaining) to give the sculpture a more uniform appearance.

A test area cleaned by laser, at an energy density of 0.4 J/cm^2, had a yellowish-brown coloration (Plate 5.13, centre) which contrasted sharply with the grey-white marble surface revealed by conventional steam cleaning. This suggested that although radiation at 1.06 μm was able to remove the particulate dirt it was unable to remove the greasy component of this layer. Tests investigating the effect of laser radiation on dirty waxed surfaces suggested that the appearance of the laser-cleaned surface was due to discoloration of this component during the cleaning process. Test areas cleaned by laser with water appeared lighter but still had a noticeable discoloration. In order to completely remove both components of the dirt layer by laser it was necessary to clean at an energy density damaging to the marble surface (Plate 5.13, left side). Tests to remove the artificial patina without damaging the marble surface were also unsuccessful. Since laser radiation at 1.06 μm discoloured rather than removed part of the dirt layer and was unable to remove the organic-based staining of the patinated surface, a combination of steam and poultice cleaning was used instead.

Cleaning of sculpture such as this presents different problems to cleaning heavily polluted outdoor sculpture due to complications arising from past interventions. In this case the laser provided the conservator with a means of removing the particulate dirt but not of removing the organic-based residues. Other techniques provided a more effective solution.

Limestone sculpture

Laser cleaning has been used successfully in a number of very different conservation projects involving limestone: to remove thick black pollution encrustations from sculpture at Lincoln Cathedral, black cement from crumbling sculptural detail, graffiti (pigment only) from ashlar surfaces and light soiling and cellulose nitrate consolidant from Egyptian reliefs. The two case studies presented here have been chosen to illustrate the versatility of the laser.

Grotesque from Lincoln Cathedral

Plate 5.14 shows a carving of a grotesque which occupied a sheltered position on the exterior of Lincoln Cathedral. The cathedral, built mainly between 1072 and 1280, is situated on a hill in the low-lying east of England. The prevailing wind direction is such that the building is exposed to emissions from the coal-fired power stations of the Trent Valley, approximately 20 km away. Dry deposition of pollutants has, in the case of the grotesque, led to the formation of a hard black crust which in some places is 15 mm thick. Cleaning has been carried out on the lower part of the front of the carving (Plate 5.15) where the crust was 4 mm thick and completely obscured the detail in the surface.

Initial cleaning was carried out with water at a relatively high energy density of 1.5 J/cm^2 and repetition rate of 8 Hz. Once most of the thickness of the crust had been removed the energy density and repetition rate were reduced to 0.4 J/cm^2 and 5 Hz respectively to allow selective removal of the crust adjacent to the carved surface. The delicate yellow-brown patina remains intact and surface detail and texture previously hidden by the crust is now visible. Cleaning of the areas covered by extremely thick black crust material has not been attempted. The most practical way of achieving this would involve the use of fine air-abrasive methods or a scalpel to remove most of the crust mechanically and then the laser to remove the crust closest to the stone surface.

An Egyptian limestone pigmented relief

Plate 5.16 shows a fourth/fifth century AD Egyptian relief excavated from Abydos in Egypt during the late nineteenth century. The relief, which measures approximately 55 × 45 cm, was carved in limestone (the flat surface around the carved figures was crudely carved with a claw chisel) and retains a large proportion of the original pigment: black pigment was used to highlight the faces of two of the figures and red pigment to highlight areas of the upper body, face, arms, feet and the lettering. The relief was covered by a thin, dark grey layer of dirt which largely obscured the carving and polychrome decoration (Plate 5.16, right side). In some areas the surface was soft and friable. Initial cleaning tests, using swabs and de-ionized water with a small amount of non-ionic detergent, resulted in a very irregular clean and left the cleaned areas with a raw appearance. Further tests were undertaken using laser radiation: at an energy density of 0.2 J/cm^2 most of the dirt was removed and the surface was left with a soft and uniform patinated appearance; at 0.4 J/cm^2 removal of dirt was more thorough but the surface appeared starker and slightly overcleaned. Since a 'soft' clean was desired cleaning was undertaken at the lower energy density.

In order to protect the pigment (black pigment is usually strongly absorbing at 1.06 μm, and previous experience has shown that certain red pigments are extremely sensitive to radiation at 1.06 μm and discolour even at low energy density (Shekede, 1997)), thin templates were accurately cut from cardboard and laid carefully onto the relief to mask the vulnerable areas, identified by microscopic examination. Cleaning of the relief, with the exception of the lettering, was carried out at 0.2 J/cm^2 using a repetition rate of 2 Hz and a beam size of 6 mm diameter (it was possible to use a relatively large spot size since the areas of pigment were obscured by templates). Plate 5.16 (left side) shows the relief during cleaning. Cleaning of the relatively complex lettering at the bottom of the relief required a different form of template. A perspex sheet (transparent to radiation at 1.06 μm) was placed over the lettering and white titanium dioxide acrylic paint (resistant to radiation at 1.06 μm) was painted onto the perspex to mask the pigment. Cleaning was then carried out through the sheet. Dirt ejected from the relief during cleaning adhered to the underside of the perspex sheet and was removed along with the sheet once cleaning had been completed. The edges of pigmented areas were then cleaned, after consolidation, working at ×10 magnification, with small cotton wool swabs dampened with de-ionized water and a non-ionic detergent. Plate 5.17 shows an area of the relief after completion of treatment; the aesthetic qualities of pigment use can be fully appreciated and the carving technique is now clearly visible.

Sandstone

Laser cleaning has been used successfully to remove compact pollution layers from external architectural detail, a hard black oxalate layer from carving on a twelfth-century doorway and relatively thin soiling layers from internal ashlar surfaces. During the course of our initial work on sandstone it became apparent that red Glasgow sandstone must be cleaned more carefully than other sandstones, otherwise the surface adopts a bleached appearance (Cooper, 1994). The colour change is believed to be due either to removal or a change in the oxidation state of the red iron oxide powder coating the silica grains of the sandstone. Care must always be taken with sandstone surfaces where the 'cement' (which binds the grains of silica together) has several components, one of which might be strongly absorbing at the wavelength being used. The case studies chosen show successful cleaning of sandstone surfaces on inside and outside building walls.

External architectural detail

Plate 5.18 shows architectural detail from the outside walls of a large nineteenth-century house in Liverpool. The porous surface of the stone was covered by a hard compact pollution layer (Plate 5.18, right side), the thickness of which varied from approximately 0.1–0.5 mm. Cleaning was carried out with water at an energy density of 0.5 J/cm^2 and a repetition rate of 10 Hz (the surface is very simple and in sound condition). The pollution layer was successfully removed from the stone leaving a uniform, clean surface (Plate 5.18, left side).

Internal ashlar

Plate 5.19 shows the cleaned surface of part of a blonde sandstone wall inside a Glasgow museum. The surface was covered by a thin, dark grey layer of dirt with a maximum thickness of approximately 0.1 mm. Cleaning was carried out with water at an energy density of 0.5 J/cm^2 and a repetition rate of 10 Hz. The rate of cleaning was measured as approximately 0.25 m^2/hr. Removal of the dirt layer revealed the patina of the blonde sandstone which gives the cleaned surface a soft mellow appearance.

Terracotta

The laser has been used successfully to clean terracotta sculpture and architectural detail. The case studies illustrate the extent to which an experienced conservator is able to control the level of cleaning with a laser.

A seventeenth-century statue of Neptune

Plate 5.20 shows a seventeenth-century overlifesized terracotta statue of Neptune, attributed to Puget, during cleaning. The surface of the statue would originally have been covered by a thin layer of terracotta wash, applied by the artist to disguise the joins between different sections (the statue was fired in several pieces). Areas of the original wash remain. In order to establish the overall condition of the statue a gentle initial clean was carried out at 0.3 J/cm^2 to remove only the black layers of dirt. Removal of the dirt layers revealed a pale yellow terracotta surface in generally sound condition (Plates 5.20, 5.21 and 5.22, right side). In areas where the dirt layers were especially thick and strongly bound to the surface a second stage of cleaning was carried out at 0.5 J/cm^2. Damage to the terracotta surface was only observed above approximately 0.7 J/cm^2. Using the laser it was possible to work through the dirt layers in such a controlled manner that areas of original terracotta wash were preserved (Plate 5.22, right side). Once cleaning had been completed conservators were able accurately to evaluate the overall condition of the statue, assess how much of the surface was covered by original pigment and undertake the appropriate conservation work.

Architectural detail

Plate 5.23 shows architectural detail from the outside of a nineteenth-century house in Liverpool. The terracotta surface was in sound condition but covered by a thin black layer of dirt (Plate 5.23, right side), a result of many years' exposure to a polluted atmosphere. Initial cleaning, carried out with water at an energy density of 0.4 J/cm^2 and a repetition rate of 10 Hz, removed the majority of dirt from the surface but failed to remove dirt from within fine cracks. A second stage of cleaning was carried out in which the conservator, working under magnification, was able to 'pick out' the remnants of dirt using a slightly higher energy density, a reduced beam size and a low repetition rate. The cleaned surface is shown in Plate 5.23 (left side). The protective fireskin of the terracotta was unaffected by the cleaning process.

Plaster

Plate 5.24 shows one of the plaster models for the Port Sunlight (Wirral, England) war memorial sculpted by William Goscombe-John in about 1920. A shellac coating, applied by the artist to seal the plaster surface, still exists and gives the surface a soft, mellow appearance. In the more complex areas, such as the head, a different casting technique had been used because of a number of undercuts and a thicker coating of dark yellow shellac had been applied. The models have, for many years, been stored in the basement of the Lady Lever Art Gallery during which time they have acquired a dark layer of dirt and dust. Using a beam size of 8 mm, a repetition rate of 10 Hz and an energy density of approximately 0.4 J/cm^2 these layers of dirt have been carefully removed from the model without any noticeable damage to either the plaster or even the shellac coating (Plate 5.24, left side). Plate 5.25 (left side) shows an area of surface with the dark shellac coating which has been cleaned and preserved. In contrast, tests carried out with mild solvents removed the shellac and left the surface with a much starker white appearance. Swabs also failed to remove dirt from within fine cracks in the surface. Careful removal of the dirt layers using laser radiation has revealed fine surface detail in the model.

Aluminium

The interaction of laser radiation with corrosion layers on metal objects is an extremely complex and, presently, little understood process. Initial work at NMGM has concentrated on the relatively simple case of cleaning corroded aluminium surfaces. The laser cleaning of more complex metal surfaces, such as bronze, is the subject of current investigations.

Figure 5.2 shows a statue of Eros by Alfred Gilbert which was cast in 1929. The statue, which is an aluminium alloy (95% aluminium, 2.5% zinc, 2.5% copper by weight), used to stand in Sefton Park, Liverpool, and is one of only two cast during the sculptor's lifetime (the other stands in Piccadilly Circus, London). Vandalism and weathering meant that to prevent further deterioration it was necessary to bring the statue indoors for conservation work. Internal armatures had corroded and the aluminium casting split, allowing moisture into the core material and migration of salts to the surface. The corroded surface consisted of a uniform dark grey layer (SEM revealed micro-cracking in crazed formation within this layer), formed through exposure of the aluminium surface to the polluted atmosphere, and other more complex white and brown deposits. Salt crystals were concentrated in areas of the surface. By using the laser very carefully, at an energy density of only approximately 0.1 J/cm^2, it was possible to remove the surface accretions and corrosion layers in an extremely controlled manner. Plate 5.26 shows a part-cleaned section of the statue. The surface on the right side has been cleaned thoroughly and left in very good condition. Cleaning was undertaken at a low energy density to avoid melting the aluminium surface. Figure 5.3 (left half) shows a partly cleaned section of a wing. The dark grey layer has been removed in such a precise and controlled way that the original tool markings have been preserved. On areas such as this cleaning was carried out using a relatively large spot size (5–7 mm) and a high repetition rate (10 Hz). On other areas where the deposits were more strongly bound to the surface a slightly increased energy density was required and cleaning was carried out with a reduced spot size (1.5–3.0 mm) and repetition rate (1–3 Hz) to avoid overcleaning. Figure 5.4 shows the statue of Eros after cleaning and repair had been completed.

Ivory and bone

Laser radiation has been used successfully to clean a number of ivory and bone objects including archaeological Egyptian artefacts and a nineteenth-century sculpture. Using a laser it is possible to clean 'thoroughly' but 'softly': dirt can be removed from within fine cracks in a surface while the surface itself retains the mellow appearance appropriate to its age.

A nineteenth-century ivory and bronze sculpture

Plate 5.27 shows a late nineteenth-century sculpture, Mors Janua Vitae, by Harry Bates. The dark winged figure is bronze and the light figure ivory. Cleaning of the large ivory figure and smaller ivory figures around the base (Plate 5.28) was carried out at an energy density of 0.8 J/cm^2 with a relatively low repetition rate of 3 Hz. The bronze figure and

Figure 5.3. *Partially cleaned section of a wing (20 cm long) from the Eros statue. The dark grey corrosion layer has been removed from the left half revealing original tool markings.*

Figure 5.2. *Statue of Eros (aluminium alloy) prior to cleaning.*

Figure 5.4. *Statue of Eros after completion of conservation work.*

bronze base (with lapis lazuli inlay) were masked, using several layers of tissue paper, to avoid accidental exposure to the laser beam. Plate 5.28 shows the ivory figures before and after cleaning. Using a relatively small beam size of 2 mm it was possible to work precisely and remove the light soiling (from small areas of the surface) without affecting other areas of the sculpture.

Archaeological artefacts

Plates 5.29 and 5.30 show ivory and bone artefacts thought to be at least 2000 years old which were excavated in Egypt at the beginning of the twentieth century. The ivory object (Plate 5.29) was cleaned at an energy density of 0.5 J/cm^2. Dirt was removed from the surface and from within fine surface cracks leaving the ivory with a uniform patinated appearance. The surface has retained its lustre. Microscopic examination of the surface at ×40 magnification failed to detect any broadening of the surface cracks or any change to the pattern of surface cracks. Dirt has been left along the top of the object to illustrate its initial soiled state. An unexpected result of the cleaning was weakening of a relatively recent (within the last 100 years) epoxy adhesive repair upon exposure to laser radiation.

Plate 5.30 shows a small piece of bone inlay, parts of which have been cleaned at different energy densities to demonstrate the extent by which the conservator is able to control the level of cleaning. Area a shows the original soiled surface. Area b was cleaned at 0.2 J/cm^2; most of the surface dirt has been removed but some dirt still remains within the fine cracks. Area d was cleaned at 0.7 J/cm^2; the surface has been cleaned more thoroughly and appears lighter. No damage to the bone surface was observed (damage visible to the naked eye occurred above approximately 1.0 J/cm^2). Area c is an intermediate stage, cleaned at 0.5 J/cm^2. The circular details in the bone surface probably contained traces of black pigment which in this case were removed by the cleaning process. Such areas of pigment could be preserved by either masking the appropriate areas or by simply cleaning around them, working under magnification with a small beam size and a steady hand.

Nineteenth-century parchment

Plate 5.31 shows a small area of script on a nineteenth-century parchment document. The document was lightly soiled (Plate 5.31, right side). Initial tests suggested cleaning should be carried out at 0.4 J/cm^2: inspection of the surface at ×40 magnification by optical microscopy and at ×10–×500 magnification by electron microscopy failed to detect any damage to the parchment at this energy density. Damage to the parchment, visible to the naked eye as lightening and physical disruption of the surface, occurred above 1.2 J/cm^2. Cleaning of the parchment (Plate 5.31, left side) was carried out at an energy density of 0.4 J/cm^2 with a beam size of 5 mm diameter and a relatively low repetition rate of 3 Hz. One pulse was sufficient to remove the complete thickness of dirt. The parchment, which was given a gentle, even clean, has maintained its mellow and aged appearance. The script has been left intact. Between five and ten pulses at 0.4 J/cm^2 were required to produce loss of ink that could be detected by the naked eye. Since only one pulse was required to clean the parchment it was possible to clean over the script without losing valuable information.

The cleaning of documents by laser is a complex process. The success of this process depends to a large extent on the type of dirt, the type of substrate, the type of ink and the way in which they are bonded together. Documents must be treated on a case-by-case basis; the energy density most appropriate for one document will almost certainly not be the most appropriate for another (the two documents might even benefit from cleaning at different wavelengths!). For example, radiation at 1.06 μm efficiently removes pencil from wood pulp paper but appears to leave a scorch mark on the paper surface (probably a result of the extremely high thermal conductivity of graphite); it has virtually no effect on ball-point pen ink at energy densities below the damage threshold of such paper. Another factor that cannot be ignored when cleaning a document by laser is the transmissivity of the substrate. A fraction of the incident laser energy will actually be transmitted through the page being worked on which could have disastrous consequences for script either on the back of the page or on the following page! Care must be taken to mask the next page and avoid dwelling on the cleaned surface. The cleaning of parchment by other types of laser and the long-term effects of laser cleaning are the subjects of more detailed research.

An oak misericord

Plate 5.32 shows an oak misericord (carved seat) which investigations revealed to be covered by six different paint and preparation layers (mainly black and brown). Tests were carried out to see whether the laser could be used to carefully work through the layers, removing them one at a time. An energy density of 0.4 J/cm^2 was required to remove the uppermost black layer. However, removal of this layer was very uneven and underlying layers were exposed to laser radiation. With the process not being self-limiting these layers were either removed or altered. The uncontrolled nature of the paint removal left the surface with a patchwork effect, showing evidence of each of the layers (Plate 5.33); in some areas five layers had been removed while in other areas only the uppermost layer had been removed. Work on polychrome surfaces is a complex process since different pigments (and binders) have different absorption properties and react in different ways upon exposure to laser radiation: some pigments have low absorptivity at 1.06 μm and are relatively stable to near infrared laser radiation; other pigments have high absorptivity and flake off; and certain pigments are extremely sensitive and discolour, e.g. exposure of vermilion to radiation at 1.06 μm causes a red to dark grey colour change even at relatively low energy density. In the case study shown here careful removal of the layers by scalpel was a far more controlled process.

The flat surface of the back of the misericord presented a much simpler problem: a thin layer of dirt on oak (Plate 5.34, left side). Tests established that cleaning was self-limiting below about 0.5 J/cm^2 (exposure to pulses of higher energy density caused damage visible to the naked eye; closer inspection through an optical microscope revealed wood fibres standing perpendicular to the surface). Plate 5.34 (right side) shows the oak surface after cleaning at 0.3 J/cm^2. The dirt layer was removed in an extremely selective manner. Examination of the cleaned surface by optical microscopy at magnifications of up to ×40 revealed no signs of damage.

Summary

The case studies presented here demonstrate the versatility of the Q-switched Nd:YAG laser as a cleaning tool. Short pulses of laser radiation in the near infrared have been used to remove thick, hard pollution encrustations from stone sculpture, thin layers of soiling from parchment and even particles of dirt from within fine surface cracks in ivory carving. Corrosion products and remnants of limewash have also been removed. The conservator has been able to clean both smooth and porous surfaces, to clean softly or thoroughly and to preserve important transition layers beneath the dirt, such as the patina. The same tool (a 300 mJ/10 Hz system) has also been used precisely to remove dirt surrounding remnants of polychromy and to clean over-lifesized sculpture.

A point that cannot be stressed often enough is that the final quality of cleaning is determined by the conservator and not by the laser. The laser is simply another tool for the conservator to use and, as with any tool, it should be used appropriately and skilfully. When used well the laser offers a high degree of precision and control, but when used badly the damage caused to the artwork will be as serious as that caused by a poorly used air-abrasive system, for example. The interaction of near infrared radiation with pigments is, at the moment, little understood and cleaning of surfaces with remnants of polychromy is usually carried out in such a way that the pigment is not exposed to the beam. Laser cleaning of paintings is carried out in an extremely controlled manner, using ultraviolet laser radiation to remove dirt and discoloured varnish layer-by-layer and diagnostic techniques to monitor and control the process. This is the subject of Chapter 6.

References

Cooper, M.I. (1994) *Laser Cleaning of Stone Sculpture*. PhD Thesis, Loughborough University

Cooper, M.I. and Larson, J.H. (1996) The use of laser cleaning to preserve patina on marble sculpture. *The Conservator*, **20**, 28–35

Cooper, M.I. and Larson, J.H. (1997) Laser cleaning of marble sculpture. *Restauratorenblatter*, Proceedings of LACONA I (Crete, October 1995), International Institute for Conservation of Historic and Artistic Works, Verlag Mayer, Vienna

Shekede, L. (1997) Lasers: a preliminary study of their potential for the cleaning and uncovering of wall paintings. *Restauratorenblatter*, Proceedings of LACONA I (Crete, October 1995), International Institute for Conservation of Historic and Artistic Works, Verlag Mayer, Vienna

6

Lasers in the conservation of painted artworks

Vassilis Zafiropulos and Costas Fotakis

Laser applications for conservation of painted artworks are relatively recent (Morgan, 1993; Fotakis, 1995; Fotakis, Hontzopoulos, Zergiotti, *et al.*, 1996; Fotakis, Zafiropulos, Anglos, *et al.*, 1997; Zergiotti, Petrakis, Zafiropulos, *et al.*, 1997). This is a particularly demanding application due to the high sensitivity of paints and binding media to light, requiring a good knowledge of the material and laser parameters involved. In addition to laser cleaning, intense interest also exists in the development and application of laser-based analytical techniques as tools for artwork diagnostics. Such techniques can be performed in-situ and are essential to artwork conservation and restoration since they provide important information regarding the physical and chemical structure of artworks. Laser-based analytical techniques can also be used for on-line monitoring of the cleaning process to safeguard against potential damage (Fotakis, Anglos, Balas, *et al.*, 1997; Gobernado-Mitre, Prieto, Zafiropulos, *et al.*, 1997; Maravelaki, Zafiropulos, Kylikoglou, *et al.*, 1997).

This chapter focuses on several key areas related to the use of lasers for conservation of painted artworks, based to a large extent on the findings of collaborative research between the Laser and Applications Division of the Foundation for Research and Technology-Hellas (FORTH) and the Conservation Department of the National Gallery of Athens. There are five major types of laser applications in painting conservation which, although interrelated, involve entirely different techniques and types of lasers. These include:

1. cleaning applications;
2. analysis of pigments and media;
3. imaging techniques of surfaces and patterns of underlying layers;
4. authentication applications;
5. diagnostics of surface or infrastructure defects.

Cleaning is important in maintaining the aesthetic aspects of the artwork and prolonging its lifetime by removing potentially damaging environmental pollutants from the surface or support material. In several demanding cleaning applications the use of lasers has proved effective. Another example is the removal of overpaintings. Analysis of the composition of pigments and media provides information about provenance and authenticity of a painting and can also be used to control the cleaning procedure and prevent damage. Laser spectroscopic techniques for non-destructive analysis are extremely powerful in this respect. Imaging techniques may serve a similar purpose by combining modern detector technology with advanced software for colour analysis or for observing underlying structures in the case of overpaintings. The use of lasers instead of conventional light sources may significantly improve the quality of information obtained. In another recent application, non-destructive holographic 'labelling' by laser microetching (Vainos, Mailis, Pissadakis, *et al.*, 1996) has interesting implications in verifying the authenticity of paintings. Finally, laser-based interferometric studies can reveal the existence of defects, e.g. stresses or detachments, and furthermore can be used to predict the future evolution of these defects. Holographic interferometry in art conservation has been used for some time (Gori, Grella and Guattari, 1974; Amadesi, D'Altorio and Paoletti, 1983; Paoletti, Schirripa, Facchini and Zanetta, 1993; Boone and Markov, 1995) and will be briefly discussed.

Laser ablation processes

Traditional methods of artwork conservation rely on mechanical or chemical techniques chosen by the conservator. These processes are difficult to control and are performed on very delicate, as well as complex, systems and therefore require extensive expertise to achieve an optimum result. For example, in painting conservation chemical solvents may penetrate the work and cause irreversible damage to the pigments and medium, while use of mechanical methods can lead to inadvertent removal of material from the work or destruction of the texture of the painted surface. A schematic cross-sectional representation indicating the multiple layers existing in a typical painting is shown in Figure 6.1. Usually, a rather thick primer or ground layer is applied to the substrate (canvas, wood etc.) and this constitutes the basis on which the work is drawn. Several paint layers are applied to the primer, each having a thickness of 20–50 μm. Finally, once the paint is dry, varnish is applied, serving as both a colour enhancing and a protecting layer. Common problems encountered in painting conservation include photochemical degradation of the varnish layer, accumulation of various kinds of pollutants on the surface, or even overpaints on the original painting. Cleaning of the artwork must be carried out in a way that will guarantee the integrity of the original work.

Laser cleaning is based on the removal of a well-defined layer of surface varnish under fully controlled conditions. In the case of paintings, contaminated surface layers of 20–300 μm thickness can be removed by excimer laser ablation with a spatial resolution of 0.1–1.0 μm per pulse as Figure 6.2a shows. Considering that a typical varnish (or paint) stroke applies 15–20 μm of material and that a typical varnish layer is 50–80 μm thick, corresponding to 3–4 strokes, it is clear that cleaning by excimer laser ablation can be a highly selective process, capable of leaving a thin protective layer of clean varnish over the painted area. Typical ablation rate data as a function of energy density obtained from model samples of aged varnish (dammar and gum lac) by using a KrF excimer laser

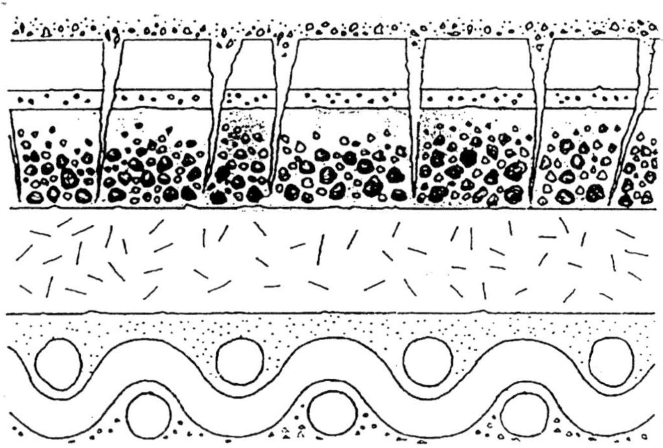

Figure 6.1. *Schematic representation of a typical cross-section through a painting.*

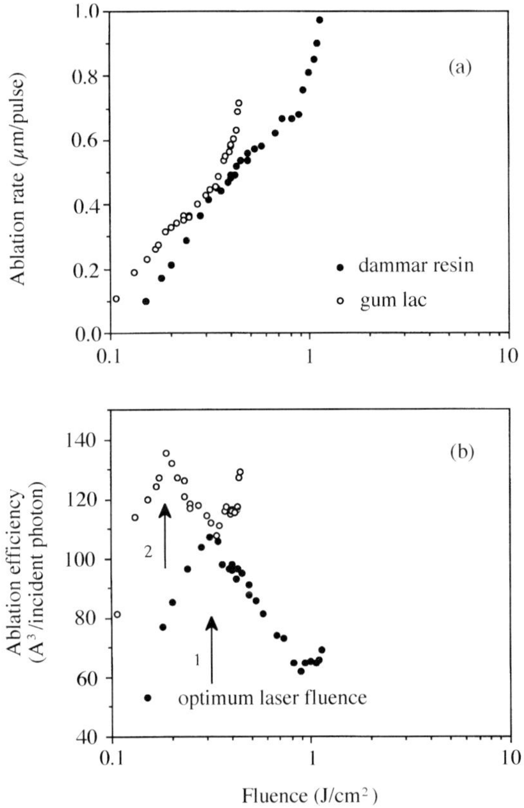

Figure 6.2. *Plots of (a) ablation rate and (b) ablation efficiency for removal of dammar and gum lac aged varnish films using a KrF excimer laser.*

(emitting at 248 nm) are presented in Figure 6.2a. The method by which such measurements are made, as well as the dependence of ablation rate on laser wavelength are described elsewhere (Zafiropulos, Petrakis and Fotakis, 1995). The ablation rate curve (Figure 6.2a) reveals the resolution per pulse on varnish removal for the selected energy density. It can be seen that the resolution attained by the laser can be as low as 0.1 μm per pulse, which is not possible using other techniques. Figure 6.2a also demonstrates the attainable control of varnish removal, i.e. it is possible to completely adjust the depth of each layer removed by appropriately focusing the laser beam. The optimum energy density can be determined for each cleaning case from the ablation efficiency curves (Figure 6.2b). This is a significant parameter in cases where large surfaces or thick varnish layers are encountered and the goal is, therefore, to remove the maximum possible volume of material per available photon. The data presented in Figure 6.2b indicate that the optimum energy density for ablation of aged dammar and gum lac is 0.3 and 0.2 J/cm^2 respectively. Also apparent in Figure 6.2b is the similarity between the two experimental trends (one curve is shifted towards higher energy densities). As the energy density is increased above a certain value (0.9 and 0.35 J/cm^2 for dammar and gum lac respectively) the ablation efficiency starts to increase for a second time. This phenomenon may be attributed to the fact that for strong laser focusing the temperature of the surface increases rapidly above the sublimation point and, therefore, a lot of material is lost. For varnish removal, work at such high energy densities is not recommended due to the possibility of thermal damage.

As mentioned in Chapter 3, cleaning by laser ablation can be described by two major mechanisms: photothermal and photochemical. Although in principle both of these processes can occur simultaneously, in practice photothermal effects are favoured when pulsed infrared lasers are used and photochemical effects are dominant with ultraviolet lasers. Ablation of material by shock wave generation can occur either photothermally or photochemically but becomes important only at high energy density.

As discussed in Chapter 3, thermal effects are primarily determined by the thermal diffusivity parameters characteristic of the material processed and the laser pulse duration (Luk'yanchuk, Bityurin, Anisimov, *et al.*, 1996; Himmelbauer, Arenholz and Bauerle, 1996). In particular, for nanosecond, pulsed excimer lasers (at 193 or 248 nm) the heat, affected zone for typical organic materials, such as varnishes used in paintings, is estimated to be comparable to the depth to which the photons penetrate. The result is that at low repetition rates no detectable temperature increase is observed on the remaining material.

The photochemical mechanism is mostly responsible for the laser ablation of organic materials when low energy density pulses from ultraviolet lasers are used (Srinivasan and Braren, 1989). The absorbed photons result in bond-breakage and for organic materials the fragments produced are predominantly molecules, radicals or ions. In parallel to the ablation process itself, the free radical formation and photo-oxidation processes could result in a variety of unwanted effects, leading to possible long-term deterioration of the work. To address the importance of these processes during laser cleaning, model samples of cinnabar pigment in linseed oil covered with dammar varnish were prepared and artificially aged. The majority of varnish was removed using different KrF excimer energy densities (based on Figure 6.2) and the remaining pigment/medium was analysed by Gas Chromatography (GC) and Mass Spectrometry (MS) in collaboration with the Department of Conservation of the Technical Educational Institute of Athens. The first results have shown absence of oxidation products when at least a thin varnish layer is left intact. However, when all the dammar is removed and the laser directly irradiates the pigment, analysis of the remaining pigment reveals various oxidation products. Further work is in progress.

The generation of a shock wave may also be important during laser ablation (Gorodetsky, Kazyaka, Melcher and Srinivasan, 1985). The contribution becomes significant when high energy density is employed and in situations where the ablated material is mostly of an inorganic nature such as pollution encrustations on marble and stone or thick overpaintings. In such cases a large percentage of material is ejected as particulates. When removing aged varnish, the shock wave mechanism appears to have a negligible effect, especially when working at low energy density. The possible long-term consequence of the shock wave is a separate issue which is currently under investigation by applying coherent optical imaging and infrared laser reflectography to detect cracks, detachments or other damage that might be caused.

Excimer laser cleaning applications

Excimer lasers have been used in several delicate and demanding medical and material processing applications. These lasers emit light pulses of about 20 ns duration at several wavelengths in the ultraviolet (mainly 193, 248, 308 nm), and are capable of removing materials in layer-by-layer increments of 0.1–1.0 μm. This facilitates the cleaning of polluted surface layers of paintings, such as dirty or aged layers of varnish, leaving the underlying material unaffected (Morgan, 1993; Fotakis, 1995; Fotakis, Hontzopoulos, Zergiotti *et al.*, 1996; Friberg, Zafiropulos and Fotakis, 1997; Fotakis, Zafiropulos, Anglos, *et al.*, 1997; Zergiotti, Petrakis, Zafiropulos, *et al.*, 1997). Successful removal of overpaintings by excimer lasers has also been demonstrated (Fotakis, Anglos, Balas, *et al.*, 1997). In this case, an overall material thickness of up to 300 μm has been removed. Plate 6.1 shows the uncovering of a seventeenth-century painting from beneath a nineteenth-century overpainting.

In contrast to traditional techniques, an excimer laser can be employed to remove unwanted surface layers in a controllable manner. With proper selection of laser parameters, detailed investigations have proved that material can be removed through ablation with minimal thermal or photochemical effects to the underlying material. Furthermore, optical techniques can be applied on-line to evaluate the cleaning process and carefully monitor its progress preventing damage to the painting (Fotakis, Anglos, Balas *et al.*, 1997; Gobernado-Mitre, Prieto, Zafiropulos *et al.*, 1997; Maravelaki, Zafiropulos, Kylikoglou, *et al.*, 1997). An example of varnish removal by KrF excimer laser (248 nm) is shown in Plate 6.2. In this case only the top layers of varnish were removed leaving behind a thin protective layer of original varnish. According to many conservators this layer also preserves the touch of time on the artwork. For comparison, two zones cleaned with the traditional method are presented at the bottom of Plate 6.2. Note that in the latter case the chemicals have removed all the varnish and the original pigments are now exposed. In the long term this may prove hazardous for the pigments since the solvents used are able to penetrate into the cracks in the pigment (Boon and Van Och, 1996). Argon Fluoride (ArF) excimer lasers (193 nm) have also been used for aged varnish removal with excellent results. The main disadvantage of the ArF laser, however, is its low ablation efficiency (Zafiropulos, Petrakis and Fotakis, 1995) which, when combined with the low output power means that the laser processing becomes extremely lengthy and expensive. Excimer lasers can also be used to remove biological debris and/or sterilize various types of paper substrates (Szczepanowska and Moomaw, 1994; Friberg, Zafiropulos, Kalaitzaki, *et al.*, 1997; Friberg, Zafiropulos and Fotakis, in press; Fotakis, Hontzopoulos, Zergiotti *et al.*, 1996). In these cases, the procedure becomes particularly demanding and complex when the paper is printed. Laser treatment in such circumstances requires advanced experimental skills.

Generally, there are three types of cleaning applications for which excimer lasers are well suited. These include the removal of *surface layers* of varnish or contaminated regions from paintings, the cleaning of a *painting's support material*, usually canvas, wood or paper, and the removal of *overpaintings*. Each of the above cases must be approached using a separate strategy in order to achieve optimum results. For instance, the laser surface cleaning of paintings ideally removes photodegraded varnish layers without any contamination or exposure of the underlying painted surface when an energy density of approximately 0.2 to 0.3 J/cm^2 is used. Removal of unwanted overpainting can also be achieved in microscopic increments, avoiding the more laborious and risky procedures which employ mechanical means or strong solvents. In this case, however, an energy density one order of magnitude greater must be employed.

For painting cleaning applications using excimer lasers, a workstation has been developed (Gobernado-Mitre, Prieto, Zafiropulos, *et al.*, 1997; Maravelaki, Zafiropulos, Kylikoglou, *et al.*, 1997; Zergiotti, Petrakis, Zafiropulos, *et al.*, 1997) comprised of:

1. an *x-y-z* stage, for precise mounting and movement of the painting;
2. suitable optics for laser beam delivery and focusing;
3. diagnostic modules for on-line monitoring and control of the cleaning process.

More specifically the *x-y-z* stage is a computer-controlled three-directional mechanical translator which is used for precise movement of the painting with respect to the laser beam. Image processing techniques or a module for broadband reflectography are used for on-line process control. As dirt is removed, a clean thin layer of varnish is exposed which can be detected by means of diffuse reflectance measurements. Alternatively, laser-based spectroscopic techniques can be incorporated for on-line control. Furthermore, various diagnostic methods such as holography, optical microscopy, profilometry and gas chromatography combined with mass spectral analysis have been used to assess the results of the cleaning process (e.g. possible oxidation of the binding medium).

Analytical and imaging applications of laser technology

Analytical techniques are vital to artwork conservation since they provide substantial information regarding the physical and chemical structure of artworks. Furthermore, profound interest also exists for adapting some well-known laser-based analytical techniques to artwork diagnostics. The main advantages of laser-based techniques over traditionally-applied techniques are the non-destructiveness and the in-situ capabilities they provide. Successful examples include the application of:

1. *Laser Induced Fluorescence* (LIF) for diagnostics of fluorescent pigments, binding media and varnishes (Miyoshi, 1988; 1990; Anglos, Solomidou, Zergiotti, *et al.*, 1996; Anglos, Couris, Mavromanolakis, *et al.*, 1997; Fotakis, Anglos and Balas, 1997; Fotakis, Zafiropulos, Anglos, *et al.*, 1997).
2. *Laser Induced Breakdown Spectroscopy* (LIBS) for pigment analysis (Fotakis, Zafiropulos, Anglos, *et al.*, 1997; Fotakis, Anglos, Couris, *et al.*, 1996; Anglos, Couris, Mavromanolakis, *et al.*, in press), layer-by-layer in-situ encrustation analysis and on-line process control during the laser cleaning of marble (Fotakis, Anglos, Balas, *et al.*, 1997), stone (Gobernado-

Mitre, Prieto, Zafiropulos, *et al.*, 1997), varnish and overpaints (Maravelaki, Zafiropulos, Kylikoglou, *et al.*, 1997).
3. *High-resolution holography* for documentation and analysis of various art objects (Gori, Grella and Guattari, 1974; Amadesi, D'Altorio and Paoletti, 1983; Paoletti, Schirripa, Facchini and Zanetta, 1993; Boone and Markov, 1995; Von Bally, 1997), damage assessment and diagnostics and evaluation of laser cleaning.
4. *Raman spectroscopy* for identification of pigments (Yúfera, Ruiz-Moreno, Manzaneda, *et al.*, 1997).

An effective application of any laser cleaning protocol depends on the availability of non-destructive diagnostic tools which include *imaging* techniques to document macroscopic colour and structural information as well as *analytical* techniques to provide information about the chemical composition of the art object. Macroscopic observations are made by reflectography and can aid on-line control of the cleaning process. Depending on the range of wavelengths used in the reflectographic observations different types of information can be obtained: reflectance spectroscopy in the ultraviolet (200–350 nm) provides information about the superficial varnished layers, while imaging in the near infrared (1.0–2.5μm) reveals underlying structural characteristics, particularly in the case of overpaintings. Illumination by modern tunable lasers incorporating optical parametric oscillators can be useful for observation of pigment layers in this respect. Coherent laser imaging methods, such as Holographic and Speckle Interferometry, have become very important in determining the presence and shape of cracks anywhere in the mass of the pigment, as well as in evaluating the results of laser cleaning. Another laser spectroscopic technique, LIF, can be used for in-situ chemical analysis while LIBS has shown promise in elemental analysis of pigments and/or contaminants as well as assisting in control of the laser cleaning process. These three latter techniques are of particular interest.

LIF is a versatile, non-destructive analytical technique, which can be performed in-situ on the artwork and provides information which can be directly related to the molecular structure of pigments or binding media, both inorganic and organic. In a typical LIF experiment a low intensity (continuous or pulsed) laser beam is used to excite a small area on the surface of the sample. Following excitation, organic or inorganic fluorescent substances in the sample emit broadband fluorescence with characteristic spectral features. Recently, LIF has been used for non-destructive pigment and media analysis (Miyoshi, 1988; 1990; Anglos, Solomidou, Zergiotti, *et al.*, 1996; Anglos, Couris, Mavromanolakis, *et al.*, 1997; Fotakis, Anglos, Balas, *et al.*, 1997; Fotakis, Zafiropulos, Anglos, *et al.*, 1997). Experimental work has shown that several commonly-used pigments and even mixtures of pigments can be identified in model samples based on their characteristic fluorescence emission. Furthermore, LIF has been used to identify the presence of photochemical oxidation in varnishes. Figure 6.3 shows the induced fluorescence when the third harmonic (355 nm) of a Q-switched Nd:YAG laser is used to excite two points of red colour on a restored painting. It is clear that the pigment used for the restoration contains a small quantity of zinc oxide that fluoresces below 400 nm.

LIBS is an analytical technique that can be used both for elemental analysis of adjacent layers and for on-line process control when combined with laser cleaning applications. It is a well-established, sensitive (especially on metals) and exceptionally selective analytical technique suitable for elemental or dimer determination that has found applications in diverse fields (Radziemski and Cremers, 1989). As the majority of encrustation or overpaint layers, as well as pigments used in paintings, are metal-containing compounds, LIBS is strongly recommended for cross-sectional analysis with high spatial resolution, defined by the ablation rate per pulse, i.e. in the submicron range. A typical quantity of pigment removed by one ultraviolet laser pulse of low energy and appropriately focused (sufficient to produce plasma) is of the order of one-tenth of a nanogram. Of course, in the case of cross-sectional analysis the quantity required increases proportionally to the thickness of the layers being analysed. This quantity

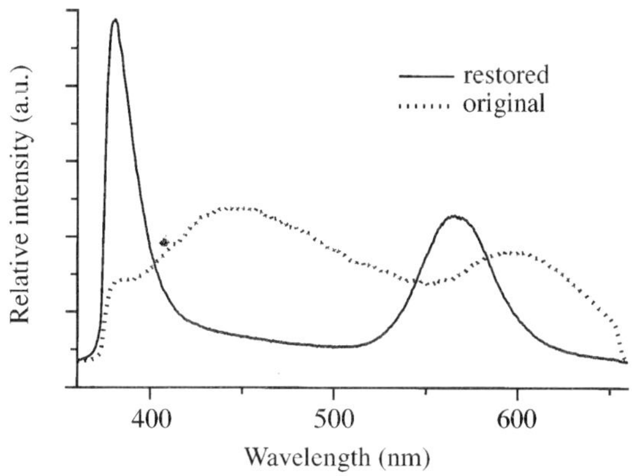

Figure 6.3. *Different LIF spectra from two points (one of them restored) of red colour on a restored painting, obtained using the third harmonic of a Q-switched Nd:YAG laser.*

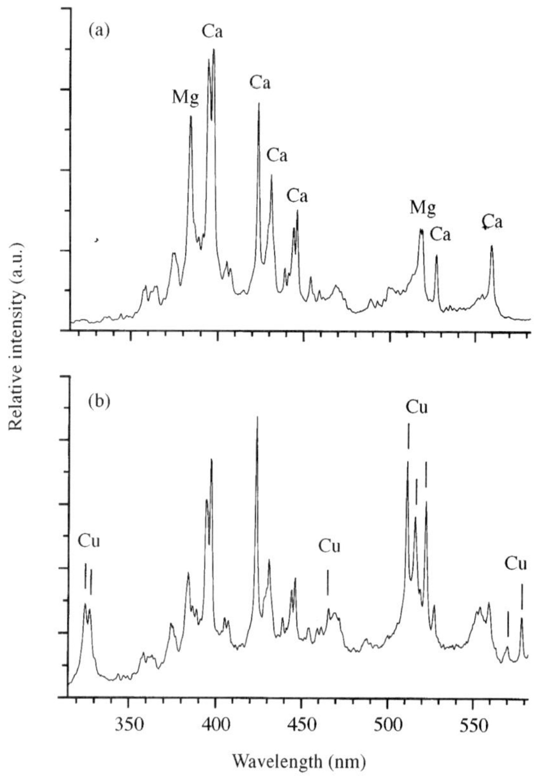

Figure 6.4. *LIBS spectra obtained from two points, a and b, of an ancient object.*

is negligible compared to the pigment lost using the traditional method of etching (a few micrograms are needed for the diamond cell Fourier Transform Infrared (FTIR) technique (Learner, 1996)), and allows extremely high spatial resolution. Furthermore, no sample preparation is required while an appropriately designed integrated LIBS system may be transportable to the place of the artwork. Figure 6.4 shows representative LIBS spectra obtained from two different points of an ancient object from the Museum of Cycladic Art. The characteristic copper emission lines in Figure 6.4b (marked with ticks) are attributed to traces of pigment containing copper, while in Figure 6.4a most of the emission lines are due to calcium and magnesium which are the predominant metallic elements of the substrate. In a LIBS experiment, during the ablation process a microscopic crater only several tenths of a micron in diameter and 0.1 μm in depth is created. As a result, the extent of damage to the artwork is barely detectable by the naked eye (especially when ultraviolet excimer lasers are used) and the technique can be characterized as practically non-intrusive. Furthermore, as the ablation depth achieved by a single nanosecond laser pulse is very small, the LIBS technique can be used to perform cross-sectional analysis of painted artworks and obtain information regarding multiple pigment layers or layers of overpaintings (Fotakis, Anglos, Balas, *et al.*, 1997). Work is in progress to investigate the experimental parameters affecting the size of the crater in order to understand how it is possible to further minimize its diameter and control its depth, especially when performing cross-sectional analysis.

Coherent laser imaging methods are another example of laser technology applied to artwork conservation (Gori, Grella and Guattari, 1974; Amadesi, D'Altorio and Paoletti, 1983; Paoletti, Schirripa, Facchini and Zanetta, 1993; Boone and Markov, 1995). Holographic Interferometry is based on comparison between the object (or its hologram) and its hologram under ambient or forced deformation conditions. The result is a three-dimensional image of the object overlaid with interference fringes, revealing cracks or detachments on the surface or inside the mass

Figure 6.5. *Holographic interferograms from two regions of the painting shown in Plate 6.2; (a) laser-cleaned (b) original uncleaned surface. Detached areas of aged varnish appear as circular patterns on the original surface. A thin sound layer of varnish remains on the laser-cleaned surface.*

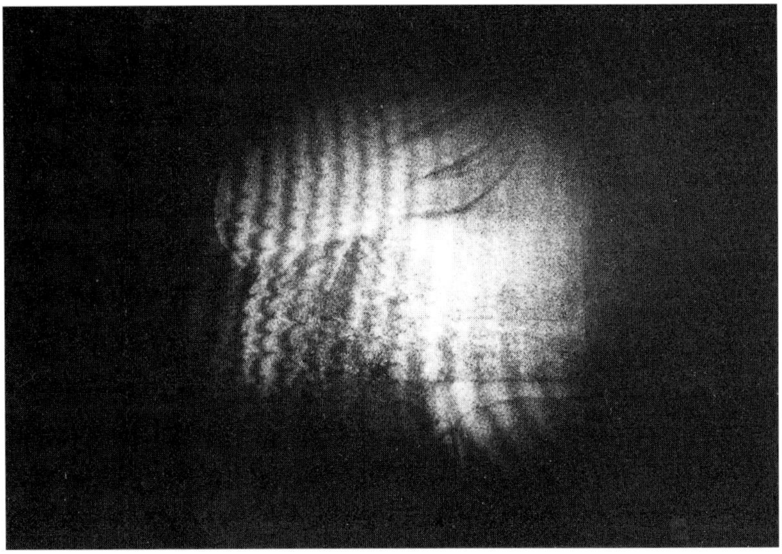

(a)

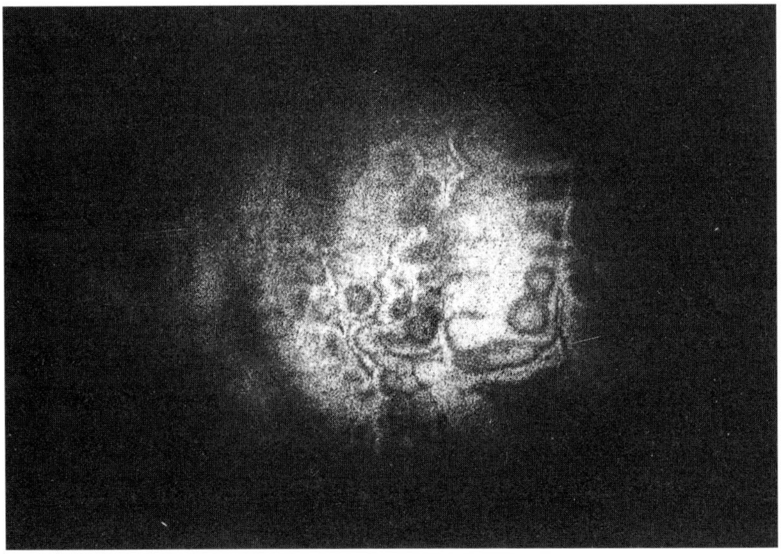

(b)

of the painting. Laser Speckle Interferometry employs interferograms produced by two speckle patterns obtained from coherent illumination of the object under test. It can detect static or dynamic deformations which are large if a crack is present below the surface. An example of using Holographic Interferometry for evaluation of laser cleaning of the painting shown in Plate 6.2 is presented in Figure 6.5. The two holographic interferograms correspond to the laser-cleaned (Figure 6.5a) and original (Figure 6.5b) surfaces. The circular patterns seen in Figure 6.5b are due to detachments between the layers of aged varnish. The interferogram for the cleaned surface shows regular interference fringes (Figure 6.5a). In this case the aged varnish layers have been removed but a thin sound varnish layer has been left on the painting. The circular patterns are no longer visible.

In summary, LIF is a rigorously non-destructive technique and although not as selective as LIBS, due to the broadband nature of molecular fluorescence spectra, has the ability to analyse both organic and inorganic materials and provide information that can be associated to their molecular structure and/or the extent of ageing. One limitation of LIF is the analysis of pigments with extremely low fluorescence quantum yield values which are practically impossible to detect, especially in the presence of small quantities of impurities which fluoresce. LIBS offers exceptional selectivity as a result of the sharp line emission spectra obtained in the analysis. The individual frequencies and relative positions of the atomic emission lines provide characteristic fingerprints for most elements which lead to their unambiguous identification. Possible reasons for diminished selectivity can be interference from background emission from the organic matrix (painting medium) in which the pigment is dispersed, similar elemental composition of different pigments or even the poor detection sensitivity of some elements (mainly non-metals). Finally, the laser interferometric methods complement the above techniques by offering a valuable tool for in-depth structural analysis of paintings and assessment of cleaning.

LIBS for on-line monitoring and control during laser ablation

Laser processing (removal of aged varnish or an overpainting) of an artwork requires a technique that can be used for on-line monitoring and control of the process. LIBS has proven to be an ideal technique for this purpose since the laser ablation process itself produces a plasma, the emission of which is characteristic of the composition of the ablated material. In most cases, the layer spectrum changes as the material removal progresses towards the final step, and it is exactly this change that can be used to guide and control the laser ablation process.

For automatic control during the laser cleaning process, LIBS spectra are continuously recorded in a pre-selected spectral region. The pre-selection is based on the results of a preliminary study that is necessary at the beginning of each project. The data acquisition processor is programmed to calculate particular peak intensity ratios that are then used to calculate the value of an algorithm designed for each specific case. When the measured values fall outside the boundaries of the algorithm, the laser stops firing and the motorized *x-y-z* stage moves the painting or the laser beam to a new position, where the cleaning process starts again. The distance that the sample is moved in each step is determined by an overlapping protocol for the sequential sample movement in both *x* and *y* directions. This is a parameter introduced in the computer program which is directly dependent on the laser parameters and on the experimental arrangement. Therefore, the algorithm, the overlapping parameters, the laser wavelength, the energy density and the pulse repetition rate are all crucial for an optimum cleaning process, and their selection must be carefully tested and optimized. Two typical examples of on-line monitoring and control based on LIBS are presented here: one refers to the removal of an overpainting and the other to the laser cleaning of aged varnish.

The possibility of removing overpaint layers from a wall plaster attributed to Jose Almada Negreiros was investigated in collaboration with the Institutes of Superior Tecnico and Jose de Figueiredo in Lisbon. A small segment, shown in Plate 6.3, was tested. Initially, an in-situ cross-sectional analysis using LIBS technique was performed on a tiny spot (hardly seen at the left corner of the sample in Plate 6.3) to identify possible changes in the composition of the overpaint layers. The laser used, an appropriately focused KrF excimer laser, was also used to remove the overpaint. Three representative LIBS spectra sampled from different depths are presented in Figure 6.6. One can see the relative change in the emission intensities of the different elements that reflects the variation in the composition of the corresponding layers; at increased depth into the overpaint the emission lines of lead, calcium and magnesium decrease while those of barium and zinc increase. Based on this result, two attempts of overpaint removal using the laser-scanning protocol described above were made (see

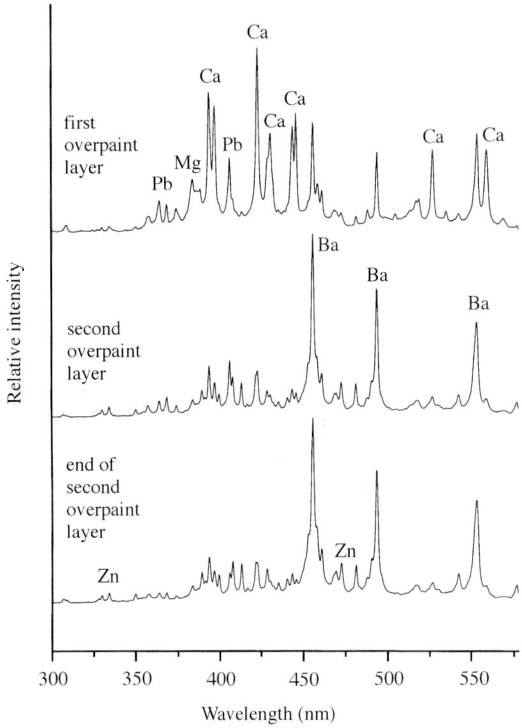

Figure 6.6. *LIBS spectra revealing the cross-sectional elemental composition of the overpaint shown in* Plate 6.3.

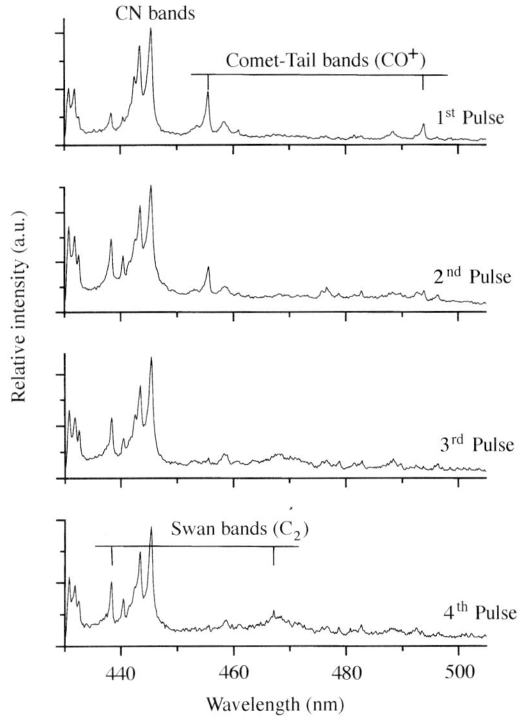

Figure 6.7. *LIBS spectra observed when successive pulses of a KrF excimer laser ablate aged varnish.*

Plate 6.3). The variables of the controlling algorithm were initially estimated from the cross-sectional data of Figure 6.6 and used to reveal the first area (square region with yellow and green in Plate 6.3). The variables were then optimized and a second scan was made with better results (see rectangular yellow portion in Plate 6.3). In both cases the laser process stopped before the exposure of the original paint and any remaining overpaint was easily removed by gentle use of a cotton tip.

Another type of monitoring application based on LIBS is the on-line control of the removal of aged varnish from paintings. Here, although the concept is the same as in overpaintings, there is a considerable difference in the origin and structure of the layers to be removed. Varnish is an organic material that is polymerized in a three-dimensional manner as its ageing progresses. In addition to polymerization, there is also natural oxidation (De la Rie, 1988) especially at the outer layers, as well as deposition of various particles (dust, soot, salts, etc.). The elemental composition throughout the varnish cross-section, however, does not change with ageing. In initial LIBS experiments on real samples of damaged paintings we found that, under certain conditions and laser parameters, there is a change in the LIBS spectrum as the varnish removal progresses towards the final varnish layers. These spectra are almost completely due to excited dimers (neutral, ionic or radicals) which are the main fragments produced in KrF excimer laser ablation of polymers (Zafiropulos, Petrakis and Fotakis, 1995). Figure 6.7 shows a typical series of LIBS spectra observed when successive pulses of a KrF excimer laser ablate aged varnish. The four spectra are identical except for two emission peaks, designated as Comet-Tail bands of CO^+, that progressively disappear. It should be emphasized that the goal is *not* to remove all the varnish but rather only the outer layers that are usually darker than the inner ones. Similar spectra have been observed in many case studies. These bands originate from the carbonyl groups (products of oxidation (De la Rie, 1988)) the density of which is expected to be higher at the outer varnish layers than at the inner ones, owing to the absorption of light mainly in the outermost layers. Plate 6.4 shows several trials of aged varnish removal with the help of on-line control, in a case study where the total depth of varnish was of the order of only a few microns. The algorithm of the process control was based on the data shown in Figure 6.7 and the basic variable used was the ratio between the peaks at 455.5 nm and 438.3 nm.

Summary

The implementation of modern laser technology in the field of art conservation has proven to have exciting possibilities and prospects both in cleaning as well as diagnostic applications. The successful cleaning of various types of artworks is dependent on the appropriate choice of laser parameters and on-line monitoring and/or controlling techniques. Such techniques are also useful for *in-situ* and *non-destructive* or *non-intrusive* analysis of pigments and media. Extreme care should be taken to optimize the laser parameters prior to the final laser application, to ensure the absence of unwanted effects on the artwork. The main difficulty of such a task originates from the fact that there are no rules for the choice of laser parameters. On the contrary, a case-by-case problem analysis is vital for assessing the possible effects that the particular choice of parameters might have. It should be emphasized that the successful implementation of lasers in art conservation is an interdisciplinary field relying on collaboration between art historians, conservators and natural scientists.

References

Amadesi, S., D'Altorio, A. and Paoletti, D. (1983) Realtime holography for microcrack detection in ancient golden paintings. *Optical Engineering*, **22**, 660–62

Anglos, D., Solomidou, M., Zergiotti, I., *et al.* (1996) Laser-induced fluorescence in artwork diagnostics: an application in pigment analysis. *Applied Spectroscopy*, **50** (10), 1331–34

Anglos, D., Couris, S., Mavromanolakis, A., *et al.* (1997) Artwork diagnostics. Laser Induced Breakdown Spectroscopy (LIBS) and Laser Induced Fluorescence (LIF). *Restauratorenblatter*, Proceedings of LACONA I (Crete, October 1995), International Institute for Conservation of Historic and Artistic Works, Verlag Mayer, Vienna

Boon, J.J. and Van Och, J. (1996) A mass spectroscopic study of the effect of varnish removal from a nineteenth century solvent-sensitive wax-oil painting. In *11th Triennial Meeting of ICOM Committee for Conservation* (J. Bridgland, ed.) pp. 197–206

Boone, M.P. and Markov, V.B. (1995) Examination of museum objects by means of video holography. *Studies in Conservation*, **40**, 103–109

De la Rie, E.R. (1988) Photochemical and thermal degradation of films of dammar resin. *Studies in Conservation*, **33**, 53–70

Fotakis, C. (1995) Lasers for art's sake! *Optics and Photonics News*, **6**, 30–35

Fotakis, C., Hontzopoulos, E., Zergiotti, I., *et al.* (1996) Laser applications in painting conservation. In *1995 AIC Paintings Speciality Group Postprints* (J.H. Gorman, ed.), American Institute for Conservation of Historic and Artistic Works, pp. 36–42

Fotakis, C., Zafiropulos, V., Anglos, D., *et al.* (1997) Lasers in art conservation. In *The Interface between Science and Conservation*, Preprints of the conference organised by the British Museum, 2–4 April, 83–90

Fotakis, C., Anglos, D., Couris, S., *et al.* (1996) Laser technology in art conservation. In *Proceedings of the Eighth International Symposium on Resonance Ionization Spectroscopy and its Applications (RIS 96)* (N. Winograd and J.E. Parks, eds) pp. 183–88, American Institute of Physics Press

Fotakis, C., Anglos, D., Balas, C., *et al.* (1997) Laser Technology in Art Conservation. In *OSA Trends in Optics and Photonics Series* (A.C. Tam, ed.), **9**, pp. 99–104, Optical Society of America, Washington DC

Friberg, T.R., Zafiropulos, V. and Fotakis, C. (1997) Removal of fungi and stains from paper substrates using laser cleaning strategies. In *Restauratorenblatter*, Proceedings of LACONA I (Crete, October 1995), International Institute for Conservation of Historic and Artistic Works, Verlag Mayer, Vienna

Friberg, T.R., Zafiropulos, V., Kalaitzaki, M., *et al.* (1997) Excimer laser cleaning of mold-contaminated paper: sterilization and air quality considerations. *Lasers in Medical Science*, **12**, 55–59

Gobernado-Mitre, I., Prieto, A.C., Zafiropulos, V., *et al.* (1997) On-line monitoring of laser cleaning of limestone by laser induced breakdown spectroscopy. *Applied Spectroscopy* (in press)

Gori, F., Grella, R. and Guattari, G. (1974) Holographic methods for painting diagnostics. *Applied Optics*, **13**, 2009–13

Gorodetsky, G., Kazyaka, T.G., Melcher, R.L. and Srinivasan, R. (1985) Calorimetric and acoustic study of ultraviolet laser ablation of polymers. *Applied Physics Letters*, **46**, 828–30

Himmelbauer, M., Arenholz, E. and Bauerle, D. (1996) Single-shot UV-laser ablation of polyimide with variable pulse lengths. *Applied Physics A*, **63**, 87–90

Learner, T. (1996) The use of FT-IR in the conservation of twentieth century paintings. *Spectroscopy Europe*, **8**, 14–19

Luk'yanchuk, B., Bityurin, N., Anisimov, S., *et al.* (1996) The role of excited species in ultraviolet-laser materials ablation. III. Non-stationary ablation of organic polymers. *Applied Physics A*, **62**, 397–401

Maravelaki, P.V., Zafiropulos, V., Kylikoglou, V., *et al.* (1997) Laser induced breakdown spectroscopy as a diagnostic technique for the laser cleaning of marble. *Spectrochimica Acta B*, **52**, 41–53

Miyoshi, T. (1988) Fluorescence from colors used for Japanese painting under N2 laser excitation. *Japanese Journal of Applied Physics*, **27**, 627–30

Miyoshi, T. (1990) Fluorescence from resins for oil painting under N_2 laser excitation. *Japanese Journal of Applied Physics*, **29**, 1727–28

Morgan, N. (1993) Excimer lasers restore 14th century icons. *Opto and Laser Europe*, **7**, September, pp. 36–37

Paoletti, D., Schirripa, S.G., Facchini, M. and Zanetta, P. (1993) Artworks diagnostics with fiber-optic digital speckle pattern interferometry. *Journal of Applied Optics*, **32**, 6236–41

Radziemski, L.J. and Cremers, D.A. (1989) Spectrochemical analysis using plasma excitation. In *Laser-Induced Plasmas and Applications* (L.J. Radziemski and D.A. Cremers, eds) pp. 295–325, Marcel Dekker

Srinivasan, R. and Braren, B. (1989) Ultraviolet laser ablation of organic polymers. *Chemical Reviews*, **89**, 1303–16

Szczepanowska, H.M. and Moomaw, W.R. (1994) Laser stain removal of fungus-induced stains from paper. *Journal of the American Institute for Conservation*, **33**, 25–32

Vainos, N.A., Mailis, S., Pissadakis, S., *et al.* (1996) Excimer laser use for microetching computer-generated holographic structures. *Applied Optics*, **35**, 6304–19

Von Bally, G. (1997) High-resolution (color) holography for documentation and analysis in cultural heritage preservation. *Restauratorenblatter*, Proceedings of LACONA I (Crete, October 1995), International Institute for Conservation of Historic and Artistic Works, Verlag Mayer, Vienna (in press)

Yúfera, J.M., Ruiz-Moreno, S., Manzaneda, M.J., *et al.* (1997) Comparative identification of the pigments in a XIXth century oil painting by means of Raman spectroscopy and electron microprobe: Goya or Lucas? *Restauratorenblatter*, Proceedings of LACONA I (Crete, October 1995), International Institute for Conservation of Historic and Artistic Works, Verlag Mayer, Vienna

Zafiropulos, V., Petrakis, J. and Fotakis, C. (1995) Photoablation of polyurethane films using UV laser pulses. *Optical and Quantum Electronics*, **27**, 1359–76

Zergiotti, I., Petrakis, A., Zafiropulos, V., *et al.* (1997) Laser applications in painting conservation. *Restauratorenblatter*, Proceedings of LACONA I (Crete, October 1995), International Institute for Conservation of Historic and Artistic Works, Verlag Mayer, Vienna

7

Future developments

Martin Cooper

Although the potential of lasers for cleaning artworks was demonstrated way back in the early 1970s lasers only started to become accepted within the conservation world twenty years later. Acceptance has taken this long because conservators have had to wait for the technology to develop sufficiently for laser cleaning to become a practical technique, and the introduction of any new technique is met with a large amount of scepticism. Quite simply, it takes many years for a new technique to become widely accepted. The introduction of laser-based techniques is now seen by many conservators as a significant advance in the refinement of conservation methods, making them more controllable and less intrusive. Careful use of a laser allows the conservator to select not only the type of surface accretions that are removed but also the amount that is removed without damaging the artwork. It is important to remember that the laser simply provides the conservator with an additional tool to use and does not replace traditional techniques. The laser should only be used if it is the most appropriate tool for the job. The quality of cleaning achieved on an artefact is still determined by the conservator. The Nd:YAG laser has been developed into a commercial cleaning system suitable for work on museum objects and sculpture and detailed carving on buildings. Development of an ultraviolet laser for work in conservation is still at a relatively early stage. Excellent results have been obtained on paintings, stained glass and parchment, but the nature of ultraviolet radiation means that diagnostic techniques are necessary to control the process. Excimer laser cleaning of artworks is not yet 'conservator-friendly'. Work is restricted to research programmes and special collaborations between institutions. In a few years' time commercial ultraviolet laser cleaning systems will probably be available, just as infrared systems are available now.

In the future, development of lasers and laser-based technologies will lead to the introduction of a wider range of cleaning systems. These systems will be smaller, more mobile and cheaper as more companies become involved in their manufacture and competition increases. The cleaning effects of laser radiation at different wavelengths are the subject of much current research and it will be interesting to see if the development of multi-wavelength systems (systems which provide a choice of wavelengths) improves the versatility of the laser as a cleaning tool. Nd:YAG lasers which provide outputs at 1.06 μm, 532 nm and have a mini dye laser attachment, providing additional outputs at visible wavelengths greater than 532 nm, are available and are currently being used in some hospitals, but further development is needed to refine the system for conservation work. It is not yet clear how beneficial the additional wavelengths will be. The 532 nm output has shown promise in removing certain organic growths from stone and also in cleaning certain types of parchment.

Laser cleaning systems are also being developed for specific applications. In some cities in England the cleaning of bronze civic sculpture has become a major problem. During the 1970s bronze sculpture was painted black in a misguided attempt to halt the corrosion process. Selective removal of this thick disfiguring layer is not possible using conventional cleaning techniques so the laser is being looked at as a possible solution. The use of lasers in large conservation projects will lead to development of a wider range of cleaning systems as the customer lays down specifications (pulse energy, repetition rate etc.) for purpose-built systems.

The question of whether laser systems can be scaled up to clean buildings is an intriguing one. Feasibility studies are comparing cleaning rates achieved by commercially-available laser systems with the cleaning rates of other techniques, such as grit blasting, to discover by how much the power output would have to be scaled up to make laser cleaning a viable alternative. The development of laser systems capable of delivering the powers necessary is probably several years away. When comparing cleaning rates of different methods for large-scale work it is important not to lose sight of the main advantages of the laser, which are the extremely high quality of cleaning, the small amounts of waste generated and the minimal disruption caused. The automated cleaning of ashlar stonework on buildings by laser is another area which is likely to receive interest in the near future. At the other end of the spectrum development of small scale cleaning systems, where the conservator works with a laser beam of a fraction of a millimetre whilst viewing the surface at high magnification, might open up further possibilities in conservation.

In the near future diagnostic techniques used to monitor cleaning at ultraviolet wavelengths might find applications in infrared laser cleaning. Such techniques will provide the conservator with real-time information about the type of material removed from the surface and the reflectivity of the surface. This information could be fed into the cleaning system and used to limit the output from the laser and prevent overcleaning.

Laser cleaning is now used routinely in a number of conservation studios across Europe. As the advantages of the technique become ever clearer this number is sure to increase. Whereas five years ago involvement in laser cleaning was restricted to small groups with access to funding, lasers and physicists with an interest in the arts (not necessarily in that order!), it is now possible simply to go out and buy or rent a system or employ a conservator with a laser to carry out a specific job. Training courses are available. Laser technology is now more accessible to conservators than ever before. The last twenty years have seen the Nd:YAG laser become accepted as a conservation tool. It will be fascinating to see what advances the next twenty years bring.

Appendix

Glossary

Absorptivity The fraction of incident energy absorbed by a surface.
Anisotropy The change of physical properties with direction.
Atom The smallest chemically divisible particle of a substance.
Brightness The power emitted from a source per unit area per unit solid angle.
Collimated Parallel.
Covalent bond A type of chemical bond formed by the sharing of electrons, commonly found in organic materials.
Density The mass per unit volume of a substance.
Diffraction The failure of light to travel in perfectly straight lines (a result of its wave nature), e.g. the spreading out of light as it emerges from a small aperture or narrow slit. A diffraction grating separates light into different wavelengths by diffraction.
Dissociation The splitting of molecules into smaller molecules or atoms.
Efficiency of a laser is the ratio of power emitted by the laser to electrical power consumed.
Electromagnetic wave A transverse wave produced by the oscillation of electric and magnetic fields, e.g. light.
Electron An elementary particle having a negative charge. A constituent of all atoms.
Ellipsoidal cavity A cavity designed so that as much of the light as possible emitted by a flashlamp falls onto the lasing medium.
Energy A measure of a system's ability to do work. It is measured in joules (J).
Energy density A measure of the concentration of the energy in a laser pulse on a surface. An important parameter for comparing the effects of different lasers, provided pulse lengths are the same. The average energy density is defined as pulse energy incident per unit area of a surface. Units are joules per square centimetre (J/cm^2). The peak energy density may be higher since the distribution of energy in a laser pulse is not usually perfectly uniform across the beam (see **Hot spot**).
Energy level Quantum theory states that the energy of an atomic system can only take certain discrete values. The allowed states of the system are called energy levels.
Excimer laser A laser in which the lasing transition occurs between the excited and ground states of the excimer. An excimer is a molecule formed by the association of one excited atom with another atom in its ground state, e.g. krypton fluoride. Radiation is emitted at ultraviolet wavelengths.
Excited state The state of a system with higher energy than its ground state.
Exponential decay The decrease of a physical quantity, y, usually with time, t, according to the negative exponential law, represented by $y = ce^{-at}$ where a, c are constants and e = 2.718.

Fireskin The dense, hard, vitrified surface of terracotta formed during firing.
Fluence See **Energy density**.
Flux A measure of the strength of a field of force (e.g. magnetic, electric) through a known area.
Frequency The number of waves passing a fixed point in one second. Units are Hertz (Hz).
Ground state energy The lowest energy state of a system.
Harmonic An integral multiple of the fundamental frequency.
Heat affected zone The volume of a body affected by heat generated as a result of absorption of laser radiation.
Holography A technique which uses the interference properties of light to record a three-dimensional image of an object.
Hot spot Region of higher than average energy within a non-uniform laser beam.
Incident beam The incoming beam.
Infrared The term used to describe invisible electromagnetic radiation with a wavelength between about 700 nm and 0.3 μm.
Insulator A poor conductor of heat, electricity or both.
Intensity Luminous flux incident on a given surface per unit area.
Interference The interaction of two waves.
Interferometry The use of interference effects to measure small distances with high precision.
Interferogram A picture of an object overlaid by a fringe pattern produced by the interference of two beams of light. The fringe pattern contains information about the physical condition of the object.
Irradiation The exposure of a substance to electromagnetic radiation.
Laser ablation The ejection of material from a solid surface as a result of irradiation by a laser beam.
Lasing medium The medium in a laser which provides the appropriate energy levels for stimulated emission. Lasing action can be achieved in gaseous, solid and liquid media.
Light The part of the electromagnetic spectrum extending from the ultraviolet to the infrared.
Linewidth A measure of the spread of wavelengths emitted by a light source.
Momentum The linear momentum of a particle is the product of its mass and velocity.
Monochromatic radiation Truly monochromatic radiation is radiation of a single wavelength. The term is usually used to describe radiation restricted to a very narrow band of wavelengths.
Nebulized water Very fine mist of water droplets.
Normal to a surface is the perpendicular to the surface.
Normal-mode A laser operated in normal-mode emits energy in pulses of relatively long duration, typically microseconds or milliseconds.
Nucleus The core of the atom, containing protons (positively charged particles) and neutrons (no charge).
Optical cavity That part of the laser which contains the lasing medium and mirrors.
Optical penetration depth Depth to which photons penetrate a material.
Order of magnitude An approximation of magnitude to within a power of ten, e.g. one order of magnitude is equivalent to approximately ten, two orders of magnitude is equivalent to approximately one hundred etc. Often used when comparing physical quantities.
Patina An alteration to the surface of an object as a result of ageing, environmental change or artificial means.
Photochemical effects Chemical effects induced in a material as a result of absorption of light.
Photon The quantum of electromagnetic radiation. The energy of a photon is inversely proportional to the wavelength of light.
Photothermal effects Thermal effects induced in a material as a result of absorption of light.
Plasma A highly ionized gas.
Population inversion A condition necessary for lasing action in which a large number of atoms are in an excited state. This is a non-equilibrium state and power must be fed into the system to maintain the inversion.
Power The power of a laser is the rate at which energy is emitted. It is measured in joules/second or watts (W). The peak power of a pulsed laser is pulse energy ÷ pulse length. The average power of a pulsed laser is pulse energy × repetition rate.

Power density The peak power per unit area incident on a surface. Units are watts per square centimetre (W/m^2). Useful when comparing the effects of different lasers (with different pulse lengths) on a surface.
Pulse energy The amount of energy in a single laser pulse.
Pulse length The duration of a single laser pulse.
Pumping source The source which provides sufficient power to the lasing medium to obtain a population inversion.
Q-switched A laser operated in Q-switched mode emits energy in pulses of relatively short duration, typically nanoseconds.
Quantum Quantum theory states that certain physical quantities can only assume discrete values. The energy of an atomic system, for example, can only change by some discrete amount called a quantum. A quantum of light is a photon.
Radical A chemically-active fragment of a molecule.
Reflection When light hits a boundary between two media, part of the light is thrown back, or reflected, into the original medium.
Reflectivity The fraction of incident energy reflected by a surface.
Refraction The bending of light as it passes into a different transparent medium as a result of a change in its speed.
Repetition rate of a laser is the number of pulses emitted in one second. The unit is the Hertz (Hz).
Reradiation Emission of radiation from a surface which has just been irradiated.
Self-limiting Cleaning is self-limiting if the removal of material from a surface stops as soon as the dirt layer has been removed, even if irradiation of the surface continues.
Shock wave An intense sound wave.
Specific heat capacity The quantity of heat required to raise the temperature of unit mass of a substance by one degree. Units are joules per kilogramme kelvin (J/kg/K).
Standing wave Stationary wave. Interference of a reflected wave with an incident wave can lead to the creation of a wave which remains stationary.
Steam laser cleaning A laser cleaning technique which involves application of a thin layer of water to the surface immediately prior to irradiation.
Stimulated emission The process by which an incoming photon of energy E triggers a transition from a higher energy level E_2 to a lower energy level E_1, with the emission of a photon of energy E, where $E = E_2 - E_1$.
Strain The change of volume and/or shape of a body as a result of applied forces.
Stress A system of forces producing strain in a body.
Thermal conductivity A measure of how well a material conducts heat. Units are joules per second metre kelvin (J/s/m/K).
Thermal diffusivity A measure of how rapidly a material will accept and conduct heat. Units are metres squared per second (m^2/s).
Thermal equilibrium The condition of a system in which there is no net exchange of heat between the components.
Thin section A cross-section which has been polished to a single crystal thickness allowing inspection by transmitted light.
Transducer A device for converting a physical parameter (e.g. sound, pressure) into an electrical signal.
Transmissivity A measure of the ability of a material to transmit radiation.
Ultrasound Acoustic waves of frequencies exceeding the limits of hearing of the human ear, i.e. 20 kHz and upwards.
Ultraviolet The term used to describe invisible electromagnetic radiation with a wavelength between about 30 nm and 300 nm.
Vaporization temperature The temperature of a liquid at which its vapour pressure equals atmospheric pressure and evaporation occurs throughout the liquid.
Wavelength The shortest distance (symbol λ) in a wave between two vibrating particles of the same phase. In a regular repeating wave pattern the wavelength is the length of one complete wave, i.e. the distance between the beginning of the wave and the nearest point where the wave begins to repeat itself.
Wave-particle duality Describes both the wave-like and particle-like nature of light.

List of symbols

Symbol		*Units*
a	Absorptivity	
c	Specific heat capacity	J/kg/K
E	Energy	J
Ec	Ablation threshold* (clean surface)	J/cm^2
Ep	Ablation threshold* (polluted surface)	J/cm^2
f	Frequency	Hz
I	Power density	$J/s/m^2$
k	Thermal diffusivity	m^2/s
K	Thermal conductivity	J/s/m/K
N_i	Population of *i*th energy level	
r	Reflectivity	
t	Time	s
T	Temperature rise	K (°C)
v	Velocity	m/s
λ	Wavelength	m
ρ	Density	kg/m^3

*SI unit is J/m^2 but in conservation the unit J/cm^2 is more convenient and more widespread.

Prefixes and multiplication factors

Prefix	*Symbol*	*Multiplication factor*
giga	G	10^9 (i.e. 1 000 000 000)
mega	M	10^6
kilo	k	10^3
hecto	h	10^2
deca	da	10
deci	d	10^{-1}
centi	c	10^{-2}
milli	m	10^{-3}
micro	µ	10^{-6}
nano	n	10^{-9} (i.e. 0.000 000 001)
pico	p	10^{-12}
femto	f	10^{-15}

Chemical symbols and acronyms

Al_2O_3	aluminium oxide
Ar	argon
ArF	argon fluoride
CO_2	carbon dioxide
Er:YAG	erbium yttrium aluminum garnet
FTIR	Fourier Transform Infrared Spectroscopy
GaAs	gallium arsenide
HeCd	helium cadmium
HeNe	helium neon
KDP	potassium dihydrogen phosphate
KrF	krypton fluoride
LAPR	Laser Assisted Particle Removal
LASER	Light Amplification by Stimulated Emission of Radiation
LIBS	Laser Induced Breakdown Spectroscopy
LIF	Laser Induced Fluorescence
LMES	Laser Microprobe Emission Spectroscopy
Nd:YAG	neodymium yttrium aluminum garnet
SEM	Scanning Electron Microscopy
XeCl	xenon chloride

Index